IEE POWER SERIES 15

Series Editors: Professor A. T. Johns
J. R. Platts

DIGITAL PROTECTION FOR POWER SYSTEMS

Other volumes in this series:

DIGITAL PROTECTION FOR POWER SYSTEMS

A. T. Johns and S. K. Salman

Peter Peregrinus Ltd. on behalf of
The Institution of Electrical Engineers

Published by: Peter Peregrinus Ltd., on behalf of the
Institution of Electrical Engineers, London, United Kingdom

Peter Peregrinus Ltd.,
The Institution of Electrical Engineers,
Michael Faraday House,
Six Hills Way, Stevenage,
Herts. SG1 2AY, United Kingdom

British Library Cataloguing in Publication Data

A CIP catalogue record for this book
is available from the British Library

ISBN 0 86341 303 X

Reprinted in paperback 1997

Printed in England by Redwood Books, Trowbridge

Contents

Chapter 1

Introduction

1.1 Historical background

Digital-based relaying was first contemplated during the late 1960s. In one of the earliest papers on the subject, Rockefeller [1] suggested that all the power system equipment in a substation could be protected using digital computers, and since that time, research in digital protection has attracted many investigators. Research activity has covered virtually every protection technique, and many novel algorithms and associated hardware implementations have emerged.

In 1971, Mann and Morrison [2] suggested a distance protection algorithm based on the prediction of the peak values of current and voltage waveforms using a sample value and its derivative. The key assumption of this approach is that the current and voltage waveforms are sinusoidal. In 1972 a similar technique, based on the first and second derivative, was developed jointly by Westinghouse and Pacific Gas & Electric Company in the USA [3, 4]. Based on this technique an experimental online system for the distance protection of a transmission line was installed in a 230 kV substation for field trial. In 1975 Makino and Miki [5] suggested using two samples to predict the peak values of the fault waveforms while Gilbert and Shovlin [6] developed an algorithm for the prediction of the peak values using three samples. In both cases the authors again assumed that the faulted current and voltage waveforms are pure sinusoids.

Other developments, which occurred at about the same time, took into account the non-sinusoidal nature of the faulted current and voltage waveforms. The investigators concentrated their efforts on extracting the fundamental components from corrupted waveforms. One approach to achieve this is based on using Fourier methods. In 1971 Ramamoorty [7] suggested extracting the desired fundamental component of voltage or current from faulted waveforms by correlating one cycle of data samples with stored samples of reference fundamental sine and cosine waves. Other workers proposed using Fourier series to determine current and voltage fundamentals [8].

Johns and Martin [9] were the first to apply the Fourier transform to both current and voltage waveforms using a data window of less than one cycle of power frequency, and current and voltage transforms were then used to calculate the measured impedance. Walsh functions were also suggested by Horton in 1975 [10]. Girgis and Brown [11] then proposed the use of Kalman filters to extract fundamental frequency components from noisy waveforms and, in 1982, this work was extended to the design of a digital distance protection [12].

Many proposed methods have emerged that make no particular assumptions as to the form of the faulted waveforms, and a number of algorithms were developed that are based on representing transmission lines by either first- or second-order differential equations. In 1970, McInnes and Morrison [13] proposed that transmission lines be modelled as a series RL circuit, which resulted in a first-order differential equation. This method was improved in 1975 by Ranjbar and Cory [14] by integrating the differential equations, to determine the resistance and the inductance of the line in such a way as to eliminate any particular harmonic and its multiples. The idea was then developed for an experimental computer relaying system by General Electric Co. in 1979 [15, 16]. In the same year Smolinski [17] proposed an algorithm which included a capacitive element within the series RL model of the line. This resulted in a second-order differential equation, which can be used to determine the line resistance R and inductance L using four samples.

The introduction of UHV lines in particular brought about new and relatively difficult problems, which required new approaches to line protection. The new methods developed were based on what are often called travelling-wave techniques. The use of such techniques has been reported since 1977 by a number of investigators, notably Tagaki *et al* [18, 19], Dommel and Michels [20], Chamia and Liberman [21], Yee and Esztergalyos [22], Johns [23, 24], Vitins [25], Crossley and MacLaren [26] and Mansour and Swift [27, 28]. The intense activity in this field has resulted in several commercial developments, particularly in the field of directional comparison protection.

1.2 Performance and operational characteristics of digital protection

It has been recognised that many benefits can be gained from the application of digital protection and these can be broadly classified under five main areas.

1.2.1 Reliability

Digital relays can be designed to regularly monitor themselves. The process of monitoring involves executing the relay software in conjunction with a prespecified data set and comparing the results with those expected from a properly functioning device. If the response turns out to be different from that expected, an error is detected and the relay initiates warning signals to the operator. This feature can be extended by programming the relay to monitor its peripherals. It should be noted that self monitoring does not in itself directly improve reliability, but it does provide a means of signifying the operational state of protection equipment. This in turn has an indirect beneficial effect on overall reliability by ensuring that the number of potential malfunctions is reduced. Reliability can be improved further by building a degree of redundancy into the hardware/software design and using different operating principles within the same relay.

1.2.2 Flexibility

Digital relays are generally more flexible than conventional devices. For example, digital relays are programmable, and this in turn makes it possible to use the same hardware for performing a variety of protection and control functions by effecting changes in the software. It is also possible that the same relay can be equipped with multiple characteristics and any revisions or modifications required by changes in the operational conditions of the system can be easily accommodated with virtually no changes in the hardware structure.

1.2.3 Operational performance

Research and field experiments have shown that, in difficult applications in particular, digital relays can be arranged to perform much better than conventional relays. This is particularly so in long distance EHV/UHV transmission lines, series/shunt-compensated lines and multi-ended circuits. It is also recognised that certain features are naturally inherent to digital relays, e.g. memory action and complex shaping of operational characteristics.

1.2.4 Cost/benefit considerations

The cost of conventional relays has continued to increase during the last two decades and the cost/benefit ratio has consequently generally increased. On the other hand, the advancement in microelectronic technology has led to a substantial reduction in the cost of digital hardware but it must be remembered that, in particular, it is the cost of the software that often dominates the overall cost. Situations exist where the cost of software for commercially developed equipment exceeds that of the hardware by at least an order of magnitude, and in consequence digitally based equipment costs more than conventional equipment.

On the other hand, high-volume digital relays, e.g. overcurrent relays, are relatively cheap because the development costs are spread across many relays and volume production allows the use of special microchip technology. Sales volumes and development costs are also important; these vary significantly according to the degree of functional complexity involved. Overall, it is true to say that the substantial improvements in performance made possible by the application of digital technology have resulted in a gradual reduction in the cost/benefit ratio for digital-protection equipment.

1.2.5 Other features and functions

With the introduction of microprocessor-based protective systems, totally new features and facilities, which have no parallel in conventional technology, have been made possible. In particular, digital relays can be programmed to provide post-fault analysis of all observed transient phenomena. This is achieved by reading out sampled data that have otherwise been acquired as part of the fault-measurement process. In addition, digital equipment monitoring both voltage and current can be programmed to compute the distance to a fault immediately after the occurrence of the fault. Such information is often extremely useful to maintenance teams in the inspection of lines following sustained fault clearance. Improved fault-location techniques are assuming major importance, particularly in long transmission lines and those subject to a relatively high incidence of contingency faults over difficult terrain. The general reduction in fault-clearance times is resulting in a reduction in visual evidence of faults, and this in turn is further fuelling a demand for accurate fault location facilities as an integral part of digital protection devices.

1.3 Basic structure of digital relays

Unlike conventional relays, a digital relay consists of two main parts: hardware and software. The type of software embedded in a relay decides not only its characteristics but its function as well, i.e. whether it is an overcurrent, differential or impedance-based measuring device. An integral and important part of the software is the algorithm, which is a set of mathematical instructions used to process input currents and/or voltages to estimate system parameters such as the RMS values of signal components, measured impedance, fundamental frequency and differential currents etc. These calculated parameters are then used to decide whether the system is sound or faulty, and consequently initiate the action necessary to isolate the faulted section.

Over the past 20 years, different types of algorithms have been developed for digital protection applications and these may be broadly classified:

(i)   Sinusoidal waveform-based algorithms, which include algorithms such as sample and first derivatives, first and second derivatives, two-sample techniques and three-sample techniques.
(ii) Fourier and Walsh-based techniques.
(iii) Least-square methods.
(iv) Solution of the differential equations of a protected-system model.
(v)   Travelling-wave-based methods.

Much of the work presented in this book will be concerned with the general principles underlying the development and implementation of the above detailed algorithms. They can be applied to the protection of generator transformers, lines, busbars, switchgear and cable circuits in both transmission and distribution systems. The range of prospective applications is vast, but the main applications to date have been in the areas of lines and transformers in transmission systems. This book will concentrate on the latter applications, but without reference to specific commercial implementations, details of which are available from manufacturers' product literature.

1.4 References

1 ROCKEFELLER, G.D.: 'Fault protection with a digital computer', *IEEE Trans.*, 1969, **PAS-88**, pp. 438–461
2 MANN, B.J., and MORRISON, I.F.: 'Digital calculation of impedance for transmission line protection', *IEEE Trans.*, 1971, **PAS-90**, 1971, pp. 270–279
3 GILCHRIST, G.B., ROCKEFELLER, G.D., and UDREN, E.A.: 'High-speed distance relaying using a digital computer, Part I: System description', *IEEE Trans.*, **PAS-91**, 1972, pp. 1235–1243
4 ROCKEFELLER, G.D., and UDREN, E.A.: 'High-speed distance relaying using a digital computer, Part II,' *ibid*, pp. 1244–1258
5 MAKINO, J., and MIKI, Y.: 'Study of operating principles and digital filters for protective relays with a digital computer', IEEE Publ. No. 75 CH0990-2 PWR, Paper C75 197 9, IEEE PES Winter Power Meeting, New York, Jan. 1975, pp. 1–8
6 GILBERT, J.G., and SHOVLIN, R.J.: 'High speed transmission line fault impedance calculation using a dedicated minicomputer', *IEEE Trans.*, 1975, **PAS-94**, 1975, pp. 544–550
7 RAMAMOORTY, M.: 'Application of digital computers to power system protection', *J. Inst. Eng. (India)*, 1972, **52**, pp. 235–238
8 McLAREN, P.G., and REDFERN, M.A.: 'Fourier-series techniques applied to distance protection', *Proc. IEE*, 1975, **122**, pp. 1295–1300
9 JOHNS, A.T., and MARTIN, M.A.: 'Fundamental digital approach to the distance protection of EHV transmission lines', *Proc. IEE*, 1978, **125**, pp. 377–384
10 HORTON, J.W.: 'The use of Walsh functions for high-speed digital relaying', IEEE PES Summer meeting, San Francisco, July 20–25, 1975, Paper A 75 582 7
11 GIRGIS, A.A., and BROWN, R.G.: 'Application of Kalman filtering in computer relaying', *IEEE Trans.*, 1981, **PAS-100**, pp. 3387–3397
12 GIRGIS, A.A.: 'A new Kalman filtering based digital distance relay', *ibid*, 1982, **PAS-101**, pp. 3471–3480
13 McINNES, A.D., and MORRISON, I.F.: 'Real time calculation of resistance and reactance for transmission line protection by digital computer', *Elec. Eng. Trans. Inst. Eng. Australia*, 1970, **EE7**, pp. 16–23
14 RANJBAR, A.M., and CORY, B.J.: 'An improved method for the digital protection of high voltage transmission lines', *IEEE Trans.* 1975, **PAS-94**, pp. 544–550
15 BREINGAN, W.D., and CHEN, M.M.: 'The laboratory investigation of a digital system for the protection of transmission lines', *IEEE Trans.* 1979, **PAS-98**, pp. 350–368
16 GALLEN, T.F., and BREINGAN, W.D.: 'A digital system for directional comparison relay', *ibid*, 1979, **PAS-98** pp 948–956
17 SMOLINSKI, W.J.: 'An algorithm for digital impedance calculation using a single PI section transmission line model', *ibid*, 1979, **PAS-98**, pp. 1546–1551
18 TAKAGI, T., BABA, J., VEMURA, K., and SAKAGUCHI, T.: 'Fault protection based on travelling wave theory, Part I: Theory', IEEE PES Summer Power meeting, Mexico City, July 1977, Paper No A 77, pp. 750–753
19 TAKAGI, T., BABA, J., VEMURA, K., and SAKAGUCHI, T.: 'Fault protection based on travelling wave theory, Part II: Sensitivity analysis and laboratory test', IEEE PES Winter meeting, New York, 1978, Paper A 78, pp. 220–226
20 DOMMEL, H.W., and MICHELS, J.M.: 'High speed relaying using travelling wave transient analysis', IEEE PES Winter meeting, New York, 1978, Paper A 78, pp. 214–219

21 CHAMIA, M., and LIBERMAN, S.: 'Ultra high speed relay for EHV/UHV transmission lines – Development, design and application', *IEEE Trans.* 1978, **PAS-97**, pp. 2104–2116
22 YEE, M.T., and ESZTERGALYOS, J.: 'Ultra high speed relay for EHV/UHV transmission lines – Installation, staged fault tests and operational experience', *ibid*, 1978, **PAS-97**, pp. 1814–1825
23 JOHNS, A.T.: 'New ultra-high speed directional comparison technique for the protection of e.h.v. transmission lines', *IEE Proc.*, 1980, **127**, pp. 228–238
24 JOHNS, A.T., MARTIN, M.A., BARKER, A., WALKER, E.P., and CROSSLEY, P.A.: 'A new approach to e.h.v. direction comparison protection using digital signal processing techniques', *IEEE Trans.*, 1986, **PWRD-1**, pp. 24–34
25 VITINS, M.: 'A fundamental concept for high speed relay', *IEEE Trans.*, 1981, **PAS-100**, pp. 163–173
26 CROSSLEY, P.A., and McLAREN, P.G.: 'Distance protection based on travelling waves', *IEEE Trans.*, 1983, **PAS-102**, pp. 2971–2982
27 MANSOUR, M.M., and SWIFT, G.W.: 'A multi-microprocessor based travelling wave relay – Theory and realization', *IEEE Trans.*, 1986, **PWRD-1**, pp. 273–279
28 MANSOUR, M.M., and SWIFT, G.W.: 'Design and testing of a multi-microprocessor travelling wave relay', *IEEE Trans.*, 1986, **PWRD-1**, pp. 74–82

Chapter 2
Mathematical background to protection algorithms

2.1 Introduction

Digital protection devices involve extensive use of numerical techniques, which are implemented in real time. Such methods are often specific to digital protection, and as such are often alien to engineers trained in the development and application of previous generations of analogue or electromagnetic-based protection devices.

To understand the principles underlying digital protection technology, it is necessary to review briefly the mathematical basis of numerical algorithms. The topics covered in this Chapter therefore include finite differences, numerical differentiation, curve fitting and smoothing, Fourier analysis, Walsh analysis, and the relationship between Fourier and Walsh coefficients. It is not the intention that the material presented should be highly rigorous in the mathematical sense, but rather that it should give a working knowledge of the numerical techniques used and thus provide a basis for the work on specific protection algorithms that is presented in later Chapters.

2.2 Finite difference techniques [1, 2, 3]

Let us assume that numerical values $f(x_k)$ of some function $y = f(x)$ are given at equally spaced values of x, such that $x_1 = x_0 + h$, $x_2 = x_0 + 2h$, $\ldots$, $x_k = x_0 + kh$. $f(x_k)$ can be obtained either analytically from some formula or by signal sampling. Therefore at $x = x_k$

$$f(x_k) = f(x_0 + kh) = f_k \qquad (2.1)$$

Using given numerical values f_k, it is possible to form difference tables containing differences up to any desired order n.

In general there are three difference functions that can be derived for a given sample set. These are forward differences, backward differences and central differences. In digital protection, such differences are commonly used in the

Table 2.1   *Forward difference table for a function* $f_k = 1/\sqrt{x_k}$

x_k	$f_k = 1/\sqrt{x_k}$	Δf_k	$\Delta^2 f_k$	$\Delta^3 f_k$	$\Delta^4 f_k$	$\Delta^5 f_k$
1	1.0000					
		−0.2929				
2	0.7071		0.1631			
		−0.1298		−0.1106		
3	0.5774		0.0525		0.0827	
		−0.0774		−0.0279		−0.0656
4	0.5000		0.0246		0.0171	
		−0.0528		−0.0108		
5	0.4472		0.0138			
		−0.0390				
6	0.4083					

process of deriving, from sampled data, approximations to the rate of change (or first difference) and second differences of relaying signals.

The forward difference functions are defined as follows:

$$\Delta f_k = f_{k+1} - f_k$$

$$\Delta^2 f_k = \Delta f_{k+1} - \Delta f_k$$

It follows that, in general, for any positive value of n, the nth order difference equation takes the form of

$$\Delta^n f_k = \Delta^{n-1} f_{k+1} - \Delta^{n-1} f_k \tag{2.2}$$

By way of illustration, Table 2.1 thus shows the difference Table for the function $f(x) = 1/\sqrt{x}$.

Backward differences are basically defined in terms of present and preceding sampled values, so that the first and second backward differences are given by

$$\nabla f_k = f_k - f_{k-1} \qquad \text{for the first backward difference}$$

and

$$\nabla^2 f_k = \nabla f_k - \nabla f_{k-1} \qquad \text{for the second}$$

The nth backward difference is thus

$$\nabla^n f_k = \nabla^{n-1} f_k - \nabla^{n-1} f_{k-1} \tag{2.3}$$

Finally, the central difference function is defined in terms of values sampled at one half of the sample interval $(h/2)$ on each side of the time at which an estimate is made. Thus, the first and second central difference functions are given by:

$$\delta f_k = f_{k+1/2} - f_{k-1/2}$$

and

$$\delta^2 f_k = \delta f_{k+1/2} - \delta f_{k-1/2}$$

The nth central difference function is thus

$$\delta^n f_k = \delta^{n-1} f_{k+1/2} - \delta^{n-1} f_{k-1/2} \qquad (2.4)$$

Since samples are not usually available at half intervals, the easiest way to implement central difference techniques is by using the samples before and after the current sample, i.e. x_{k-1} and x_{k+1}. Thus, for example, the first central difference becomes

$$\delta f_k = \delta f_{k+1} - \delta f_{k-1}$$

Given a sampled data set taken at $x_{k-1}, x_k, x_{k+1}, \ldots$, the sequences of numbers defining the forward and backward difference functions are identical. The distinction between the three functions essentially lies in the time at which a given function is available for use in further digital processing and, where a difference function is used in approximating the differential of an actual signal (to whatever order), the degree of the resulting approximation. Thus, for example, the forward difference function is available at the time the $k+1$th sample is available whereas that for the backward approximation is available at the kth sample instant. Furthermore, the degree of accuracy to which a particular function can be used to approximate, say, the first differential of a signal varies according to the waveform involved. For example, with reference to Figure 2.1, it will be evident that a central difference function using samples taken at instants corresponding to $k+1/2$, $k-1/2$ may give a better approximation to the differential of the particular signal shown than is the case for the forward or backward difference function. In other words, the slope of a straight line joining BD in Figure 2.1, which corresponds to the first derivative calculated using the central difference formula, is close to the slope of the tangent at C.

In practice, it is not generally possible to assume a particular form to the waveforms encountered and the accuracy in approximation that can be achieved with each method varies from point to point as a given waveform is tracked. Each method therefore has advantages in particular applications. In any event, the basic difference functions form the basis of more accurate approximations to signal derivations by using interpolation formulas explained in Section 2.4.

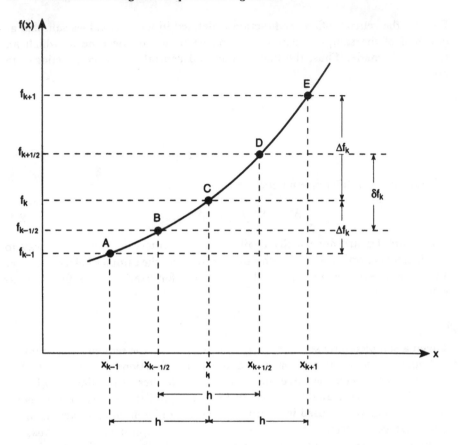

Figure 2.1   Forward, backward and centre difference function related to an arbitrary waveform

Closely associated with difference function operators Δ, ∇ and δ are the shift operator E and the average operator μ. The shift operator is defined as the operator that increases the instant at which a function is sampled by one tabular interval. Thus

$$E(f_k) = f(x_k + h) \qquad (2.5)$$

A relation between the shift operator and finite difference operators can be derived as follows:

From the definition of Δf_k, we have

$$\Delta f_k = f_{k+1} - f_k = E f_k - f_k = (E - 1) f_k$$

Therefore

$$\Delta = E - 1 \tag{2.6}$$

or

$$E = 1 + \Delta \tag{2.7}$$

Also from the definition of backward difference operator (∇) we obtain

$$\nabla f_k = (E - 1) f_{k-1} \tag{2.8}$$

from which:

$$\nabla = 1 - E^{-1} \tag{2.9}$$

The first central difference equation

$$\delta f_k = f_{k+1/2} - f_{k-1/2}$$

can now be written in terms of the shift operator:

$$\delta f_k = E^{1/2} f_k - E^{-1/2} f_k = (E^{1/2} - E^{-1/2}) f_k \tag{2.10}$$

It is thus apparent from eqn. 2.10 that the centre difference operator can be written in terms of the shift operator:

$$\delta = E^{1/2} - E^{-1/2} \tag{2.11}$$

The average operator μ is defined as follows:

$$\mu f_k = 1/2(f_{k+1/2} + f_{k-1/2}) \tag{2.12}$$

In effect, the average operator permits an estimate of a given function to be obtained from sampled values taken at half-sample intervals on each side of the time at which an estimate is required.

Using the definition of the operator E, the equation thus becomes:

$$\mu f_k = (E^{1/2} f_k + E^{-1/2} f_k)/2$$

$$= 1/2(E^{1/2} + E^{-1/2}) f_k$$

or

$$\mu = 1/2(E^{1/2} + E^{-1/2}) \tag{2.13}$$

The various operators detailed above represent a succinct way of representing sampled values. They also, importantly, allow easy manipulation of the sampled data used in algorithmic equations and are consequently widely used in the development of software for digital protection systems.

2.3 Interpolation formulas [1, 2, 3]

Different methods can be used to derive interpolating formulas. A very common approach, which is used here, is based on the previously explained shift operator.

2.3.1 Forward interpolation

Given a function such as that shown in Figure 2.1, its forward value $f_{k+1} = f(x_k + h)$ can be found from its current value $f_k = f(x_k)$ using eqn. 2.5, such that

$$f_{k+1} = f(x_k + h)$$

$$= E f_k$$

Similarly, the value of the function at $x = x_k + ph$ is equal to its value at x_k shifted by an interval equal to p times the sampling interval h, so that

$$f_{k+p} = E^p f_k \tag{2.14}$$

Substituting the value of E from eqn. 2.7 into eqn. 2.14 we obtain

$$f_{k+p} = (1 + \Delta)^p f_k$$

$$= [1 + k_{1f}\Delta + k_{2f}\Delta^2 + k_{3f}\Delta^3 \ldots + k_{pf}\Delta^p] f_k \tag{2.15}$$

where any value k_{mf} is often known as the mth binomial coefficient, given by

$$k_{mf} = \frac{p(p-1)(p-2) \ldots (p-m+1)}{m!}$$

It follows that an estimate of the value of a function can be obtained from the summation of terms given in eqn 2.16:

$$f_{k+p} = f_k + p\Delta f_k + \frac{p(p-1)}{2!}\Delta^2 f_k + \frac{p(p-1)(p-2)}{3!}\Delta^3 f_k + \cdots$$

$$+ \frac{p(p-1)(p-2) \ldots (p-n+1)}{n!}\Delta^n f_k \tag{2.16}$$

where n is the order of the polynomial.

If p is equal to r such that $0 \leqslant r \leqslant 1$, then $f_{k+p} = f_{k+r}$ is simply equal to the value of the function at $x_k + rh$, and it follows that

$$f_{k+r} = f(x_k + rh) = f_k + r\Delta f_k + \frac{r(r-1)}{2!} \Delta^2 f_k + \frac{r(r-1)(r-2)}{3!} \Delta^3 f_k + \cdots$$

$$+ \frac{r(r-1)(r-2) \ldots (r-n+1)}{n!} \Delta^n f_k \qquad (2.17)$$

Eqn. 2.17 is known as the Gregory–Newton forward interpolation formula, which is useful in determining the value of a function from its forward difference table.

Example

Calculate the value of the function $f(x) = 1/\sqrt{x}$ at $x = 3.5$ from its forward difference Table 2.1, using eqn. 2.17.

Solution

Let us first estimate the value using only the first two terms of eqn. 2.17 so that we have $f_{k+r} \approx f_k + r\Delta f_k$, where $k = 3$ and $r = 0.5$.

With reference to Table 2.1, f_3 and Δf_3 are equal to 0.5774 and -0.0774, respectively, so that

$$f_{3.5} \approx 0.5774 + 0.5(-0.0774) = 0.5387$$

This corresponds to an error of 0.786% from the exact value $f(3.5) = 1/\sqrt{3.5}$, which equals 0.5345. Now let us repeat the calculations using the first three terms of eqn. 2.17, which gives

$$f_{k+r} \approx f_k + r \cdot \Delta f_k + \frac{r(r-1)}{2!} \Delta^2 f_k$$

Again, from Table 2.1, $\Delta^2 f_3 = 0.0246$, so

$$f_{3.5} \approx 0.5774 + 0.5(-0.0774) + \frac{0.5(0.5-1)}{2}(0.0246)$$

$$= 0.535625$$

This corresponds to an error of 0.21%, which indicates that the more terms of eqn. 2.17 are used, the better the accuracy obtained. However, for this example, it is sufficient to take only the first two terms. In digital protection applications, eqn. 2.17 can be very useful in determining a numerical estimate of the differential of a function; this will be discussed in Section 2.4.

2.3.2 Backward interpolation

Consider the function shown in Figure 2.1; a backward value of this function, say $f_{k-1} = f(x_k - h)$, can be used to obtain its current value $f_k = f(x_k)$ by using the shift operator E. Thus $f_{k-1} = f(x_k - h) = E^{-1} f(x_k)$, or $f_k = E f_{k-1}$.

In general the current value f_k can be found from the function f_{k-p}, which is p intervals behind it, which can be succinctly stated mathematically by $f_k = E^p f_{k-p}$.

It is often more convenient to rename the function value f_{k-p} as f_N and therefore the current value f_k becomes f_{N+p}. Thus

$$f_{N+p} = E^p f_N \tag{2.18}$$

Substituting eqn. 2.9 into eqn. 2.18, we obtain

$$f_{N+p} = (1 - \nabla)^{-p} f_N$$

or

$$f_{N+p} = [1 - k_{1b}\nabla + k_{2b}\nabla^2 - k_{3b}\nabla^3 + \cdots + k_{pb}\nabla^p] f_N$$

The binomial coefficients $k_{1b}, k_{2b} \ldots$ take the form

$$k_{mb} = \frac{p(p+1)(p+2) \ldots (p+m-1)}{m!}$$

It follows that the value of a function can be estimated by eqn. 2.19 by using the backward difference operator

$$f_{N+p} = f_N + p\nabla f_N + \frac{p(p+1)}{2!}\nabla^2 f_N + \frac{p(p+1)(p+2)}{3!}\nabla^3 f_N + \cdots$$

$$+ \frac{p(p+1)(p+2) \ldots (p+n-1)}{n!}\nabla^n f_N. \tag{2.19}$$

As previously, eqn. 2.19 can be written in terms of a function of the shift interval h so that, for a fractional shift sh say, eqn. 2.19 becomes:

$$f_{N+s} = f(x_N + sh) = f_N + s\nabla f_N + \frac{s(s+1)}{2!}\nabla^2 f_N + \frac{s(s+1)(s+2)}{3!}\nabla^3 f_N + \cdots$$

$$+ \frac{s(s+1)(s+2) \ldots (s+n-1)}{n!}\nabla^n f_N \tag{2.20}$$

This is known as the Gregory–Newton backward interpolation formula. Again, as we shall see later in this Chapter, this formula is useful in determining an estimate of the differential of a function.

2.3.3 *Central difference interpolation*

There are a number of different versions of interpolation formulas that can be obtained by using central differences [2]. Four of the more common formulas are given here for reference and completeness:

(i)   Forward Newton–Gauss interpolation formula

$$f(x_k + rh) = f_k + r\delta f_{k+1/2} + \frac{r(r-1)}{2!}\delta^2 f_k + \frac{(r+1)(r-1)}{3!}\delta^3 f_{k+1/2}$$

$$+ \frac{(r+1)r(r-1)(r-2)}{4!}\delta^4 f_{k+1/2} + \cdots \qquad (2.21)$$

(ii)   Backward Newton–Gauss interpolation formula

$$f(x_k + rh) = f_k + r\delta f_{k-1/2} + \frac{r(r-1)}{2!}\delta^2 f_k + \frac{(r+1)r(r-1)}{3!}\delta^3 f_{k-1/2}$$

$$+ \frac{(r+2)(r+1)r(r-1)}{4!}\delta^4 f_k + \cdots \qquad (2.22)$$

(iii)   Stirling interpolation formula

$$f(x_k + rh) = f_k + \frac{r}{1!}\frac{(f_{k+1/2} + f_{k-1/2})}{2} + \frac{r^2}{2!}\delta^2 f_k$$

$$+ \frac{r(r^2 - 1)}{3!}\frac{(\delta^3 f_{k+1/2} + \delta^3 f_{k-1/2})}{2} + \frac{r^2(r^2 - 1)}{4!}\delta^4 f_k + \cdots \quad (2.23)$$

(iv)   Laplace–Everett interpolation formula

$$f(x_k + rh) = rf_k + \frac{r(r^2 - 1)}{3!}\delta^2 f_k + \frac{r(r^2 - 1)(r^2 - 4)}{5!}\delta^4 f_k + \cdots$$

$$+ rf_{k+1} + \frac{r(r^2 - 1)}{3!}\delta^2 f_{k+1} + \frac{r(r^2 - 1)(r^2 - 4)}{3!}\delta^4 f_{k+1} + \cdots \quad (2.24)$$

2.4 Numerical differentiation [3, 4, 5]

Any of the interpolation formulas derived in the previous Section can be used to estimate the derivative of a sampled function. For example, let us consider the forward interpolation formula of eqn. 2.17.

$$f(x_k + rh) = f_k + r\Delta f_k + \frac{r(r-1)}{2!}\Delta^2 f_k + \frac{r(r-1)(r-2)}{3!}\Delta^3 f_k + \cdots$$

By differentiating with respect to r, we obtain

$$hf'(x_k + rh) = \Delta f_k + \frac{2r-1}{2}\Delta^2 f_k + \frac{3r^2-6r+2}{6}\Delta^3 f_k$$

$$+ \frac{2r^3 - 9r^2 + 11r - 3}{12}\Delta^4 f_k + \cdots$$

and again

$$h^2 f''(x_k + rh) = \Delta^2 f_k + (r-1)\Delta^3 f_k + \frac{6r^2 - 18r + 11}{12}\Delta^4 f_k + \cdots$$

Derivatives at the point (x_k, f_k) can be obtained by setting $r=0$ so that for the first derivative we obtain

$$f'_k = f'(x_k) = \frac{1}{h}\left(\Delta - \frac{1}{2}\Delta^2 + \frac{1}{3}\Delta^3 - \cdots\right)f_k \qquad (2.25)$$

The second derivative formula will likewise be

$$f''_k = f''(x_k) = \frac{1}{h^2}\left(\Delta^2 - \Delta^3 + \frac{11}{12}\Delta^4 - \cdots\right)f_k \qquad (2.26)$$

Similarly, from the backward interpolation eqn. 2.20 by differentiating with respect to s, we obtain

$$hf'(x_N + sh) = \nabla f_N + \frac{2s+1}{2}\nabla^2 f_N + \frac{3s^2+6s+2}{6}\nabla^3 f_N$$

$$+ \frac{2s^3 + 9s^2 + 11s + 3}{12}\nabla^4 f_N + \cdots$$

$$h^2 f''(x_N + sh) = \nabla^2 f_N + (s+1)\nabla^3 f_N + \frac{6s^2 + 18s + 11}{12}\nabla^4 f_N + \cdots$$

Setting $s=0$ leads to the first derivative, which becomes

$$f'_k = f'(x_{N=k}) = \frac{1}{h}\left(\nabla + \frac{1}{2}\nabla^2 + \frac{1}{3}\nabla^3 + \cdots\right)f_k \qquad (2.27)$$

and

$$f''_k = f''(x_k) = \frac{1}{h^2}\left(\nabla^2 + \nabla^3 + \frac{11}{12}\nabla^4 + \frac{5}{6}\nabla^5 + \cdots\right)f_k \qquad (2.28)$$

Derivatives can alternatively be calculated from central differences using any of the equations from 2.21 to 2.24; a useful central difference form is given in eqn. 2.29:

$$f'_k = \left(\mu\delta - \frac{1}{6}\mu\delta^3 + \frac{1}{30}\mu\delta^3 - \cdots \right) f_k / h$$

(2.29)

Example

Use the backward interpolating formula to derive an expression for estimating the derivative of a sampled function after

(i)   two samples are made available
(ii) three samples are made available, but the derivative is to be calculated for the instant in time that the second sample is available
(iii) repeat (ii) for the second derivative.

Solution

(i)   By considering only the first term of eqn. 2.27, we obtain an estimate based on only two samples; this is given in eqn. 2.30.

$$f'_k = \frac{1}{h}\nabla f_k = \frac{1}{h}(f_k - f_{k-1})$$

(2.30)

(ii) Since the derivative is required to be found following the second sample, let us assume that the last sample is f_{k+1}, and hence the second sample would be f_k. Now eqn. 2.27 can be expressed in terms of f_{k+1} by using the relationship $f_k = E^{-1}f_{k+1}$ which, from eqn. 2.9, can be written in the form given in eqn. 2.31:

$$f_k = (1 - \nabla)f_{k+1}$$

(2.31)

By substituting eqn. 2.31 into eqn. 2.27, and simplifying we obtain

$$f'_k = \frac{1}{h}\left(\nabla - \frac{1}{2}\nabla^2 - \frac{1}{6}\nabla^3 - \cdots \right) f_{k+1}$$

(2.32)

Now by considering only the first and second terms, eqn. 2.32 reduces to eqn. 2.33, which in turn defines, as required, an estimate of the rate of change of the function at an instant in time midway between the pair of samples used.

$$f'_k = \frac{1}{2h}(f_{k+1} - f_{k-1})$$

(2.33)

(iii) By substituting eqn. 2.31 into eqn. 2.28 we obtain

$$f''_k = \frac{1}{h^2}\left(\nabla^2 - \frac{1}{12}\nabla^4 - \frac{1}{12}\nabla^5 - \cdots \right) f_{k+1}$$

Thus by considering only the first term we obtain

$$f''_k \simeq \frac{1}{h^2}\nabla^2 f_{k+1}$$

$$= \frac{1}{h^2}(f_{k+1} - 2f_k + f_{k-1}) \tag{2.34}$$

As required, eqn. 2.34 defines a measured approximation to the second rate of change of a function at the time when the second sample is taken.

2.5 Curve fitting and smoothing [3–5]

The process of representing a data set by a mathematical function (or expressing a complicated mathematical expression by a simpler one) is known as curve fitting. A widely used procedure to achieve this involves the method of least squares, which can be used for both tabular data and functions.

2.5.1 Least squares method

Consider a set of N measured points defined by pairs of numbers:

$$(x_1, y_1), (x_2, y_2), \ldots, (x_N, y_N)$$

where x_i is the value of the independent variable taken at the ith measurement and y_i its corresponding dependent variable. Such data may be generated, for example, by the measurement of current flowing through an electrical element (independent variable) and its corresponding voltage across the element (dependent variable). In digital protection, it is common to sample a relaying voltage at discrete instants in time; in this case, the dependent variable can be regarded as the sampled value and the independent variable the discrete value of time at which it is sampled. Let us now determine a function $u(x)$ to approximate the actual function such that

$$y_i \approx u(x_i) \qquad i = 1, 2, \ldots, N \tag{2.35}$$

The type of function best used in the approximation process—polynomial, exponential etc.—depends on the nature of the problem. In most protection cases the most widely used and suitable approximating function is a polynomial of the form

$$u = a_0 + a_1 x + a_2 x^2 + \cdots + a_m x^m \qquad (m < N) \tag{2.36}$$

To define this function fully, it is necessary to choose the values of the constants $a_0, a_1, \ldots, a_m$ such that the function best fits the given data. One criterion, which is often regarded as giving the best fit, requires that the sum of the squares of the displacements of the measured points (y_i) from the curve defined by eqn. 2.36 should be a minimum.

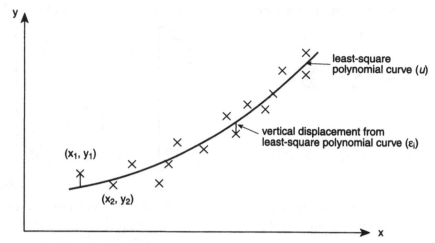

Figure 2.2 Method of least squares

Figure 2.2 illustrates the basis of curve fitting using the least-squares method. Let the vertical displacement of measured points (y_i) from the fitted curve u be ε_i, in which case:

$$\varepsilon_i = y_i - u(x_i) \tag{2.37}$$

The sum of the squares of errors is thus as given by eqn 2.38.

$$S = \sum_{i=1}^{N} \varepsilon_i^2 = \sum_{i=1}^{N} [y_i - (a_0 + a_1 x_i + a_2 x_i^2 + \cdots + a_m x_i^m)]^2 \tag{2.38}$$

The necessary condition for S to be a minimum is that its partial derivative with respect to the parameters a_0, a_2, ... etc. should be zero, i.e.

$$\frac{\partial S}{\partial a_k} = 0 \qquad (k = 0, 1, 2, \ldots, m) \tag{2.39}$$

When eqn. 2.39 is applied to eqn. 2.38, N equations will be formed, often called 'normal equations'. Thus

$$\frac{\partial S}{\partial a_0} = -2 \sum_{i=1}^{N} [y_i - (a_0 + a_1 x_i + a_2 x_i^2 + \cdots + a_m x_i^m)] = 0$$

$$\frac{\partial S}{\partial a_1} = -2 \sum_{i=1}^{N} x_i [y_i - (a_0 + a_1 x_i + a_2 x_i^2 + \cdots + a_m x_i^m)] = 0$$

$$\vdots$$

$$\frac{\partial S}{\partial a_m} = -2 \sum_{i=1}^{N} x_i^m [y_i - (a_0 + a_1 x_i + a_2 x_i^2 + \cdots + a_m x_i^m)] = 0$$

After simplification, the resulting equations in a matrix form would be

$$
\begin{bmatrix}
N & \sum x_i & \sum x_i^2 & \cdots\cdots\cdots & \sum x_i^m \\
\sum x_i & \sum x_i^2 & \sum x_i^3 & \cdots\cdots\cdots & \sum x_i^{m+1} \\
\cdots\cdots\cdots\cdots\cdots\cdots\cdots\cdots\cdots \\
\cdots\cdots\cdots\cdots\cdots\cdots\cdots\cdots\cdots \\
\sum x_i^m & \sum x_i^{m+1} & \sum x_i^{m+2} & \cdots\cdots\cdots & \sum x_i^{2m}
\end{bmatrix}
\begin{bmatrix}
a_0 \\ a_1 \\ \cdots \\ \cdots \\ a_m
\end{bmatrix}
=
\begin{bmatrix}
\sum y_i \\ \sum x_i y_i \\ \cdots \\ \cdots \\ \sum x_i^m y_i
\end{bmatrix}
$$

$$(2.40)$$

Eqn. 2.40 represents a system of simultaneous equations which can be solved for the unknown $a_0, a_1, \ldots . a_m$.

Example
The following data were obtained experimentally:

x	−1.0	−0.1	0.2	1.0
y	1.0	1.099	0.808	1.0

Use the least square method to fit a straight line to this data.

Solution
The straight line equation is given by

$$u = a_0 + a_1 x$$

The normal equations for this function are obtained directly from eqn. 2.40. Setting $m = 1$, as required for a straight line approximation, the normal matrix equation is:

$$
\begin{bmatrix}
N & \sum x_i \\
\sum x_i & \sum x_i^2
\end{bmatrix}
\begin{bmatrix}
a_0 \\ a_1
\end{bmatrix}
=
\begin{bmatrix}
\sum y_i \\ \sum x_i y_i
\end{bmatrix}
$$

$$(2.41)$$

Since there are four data points, the limit on the summations is $N = 4$, and this gives the following constant values in eqn. 2.41:

$$N = 4$$

$$\sum x_i = -1.0 - 0.1 + 0.2 + 1.0 = 0.1$$

$$\sum x_i^2 = (-1.0)^2 + (-0.1)^2 + (0.2)^2 + (1.0)^2 = 2.05$$

$$\sum y_i = 1.0 + 1.099 + 0.808 + 1.0 = 3.907$$

$$\sum x_i y_i = (-1.0)(1.0) + (-0.1)(1.099) + (0.2)(0.808) + (1.0)(1.0) = 0.0517$$

Substituting these values into eqn. 2.41 and solving for a_0 and a_1, we obtain

$$a_0 = 0.9773 \qquad a_1 = -0.0224$$

The straight line equation which best fits the data set is therefore given by

$$u = 0.9773 - 0.0224\, x$$

In the previous analysis, it has been assumed that the variable y depends only on a single variable x. The general case is such that the variable y may depend on more than a single variable. An example that illustrates this case is a three-phase unbalanced system in which the phase voltage v_a depends on the three associated phase currents i_a, i_b and i_c. Therefore, in order to determine a value of v_a, the currents i_a, i_b and i_c must be measured. In this case, the voltage v_a is a function $v_a = f(i_a, i_b, i_c)$. Thus each measured voltage is related to the measured currents by a data set $(i_{a1}, i_{b1}, i_{c1}, v_{a1})$, $(i_{a2}, i_{b2}, i_{c2}, v_{a2}) \ldots (i_{aN}, i_{bN}, i_{cN}, v_{cN})$. More generally, for a function of z variables, the data set involved would take the form $(x_{11}, x_{21}, x_{31}, \ldots x_{z1}, y_1)$, $(x_{12}, x_{22}, x_{32} \ldots x_{z2}, y_2)$, $\ldots$ $(x_{1N}, x_{2N}, x_{3N}, \ldots x_{zN}, y_N)$. The principles involved in deriving a least-squares-error based function for extended data sets associated with multivariable functions are similar to those outlined above for the single-variable case. Full details of the precise methodology involved can be found in References 4 and 5.

2.5.1.1 Least squares and pseudoinverse [6–9]
The solution of linear equations having unknowns less than the number of equations resembles a situation similar to the least-squares method. For example, let us consider the following n linear equations having m unknowns, where $n > m$:

$$a_{11}x_1 + a_{12}x_2 + \cdots + a_{1m}x_m = y_1$$

$$a_{21}x_1 + a_{22}x_2 + \cdots + a_{2m}x_m = y_2$$

$$\cdots \cdots \cdots \cdots \cdots \cdots \cdots \cdots \tag{2.42}$$

$$a_{n1}x_1 + a_{n2}x_2 + \cdots + a_{nm}x_m = y_n$$

Writing these equations in a more succinct way results in

$$\underset{n \times m}{A} \quad \underset{m \times 1}{X} = \underset{n \times 1}{Y} \tag{2.43}$$

where

$$
A = \begin{bmatrix}
a_{11}\ a_{12}\ \cdot\cdot\cdot\cdot\cdot\cdot\ a_{1m} \\
a_{21}\ a_{22}\ \cdot\cdot\cdot\cdot\cdot\cdot\ a_{2m} \\
\cdot\cdot\cdot\cdot\cdot\cdot\cdot\cdot\cdot\cdot\cdot\cdot\cdot \\
\cdot\cdot\cdot\cdot\cdot\cdot\cdot\cdot\cdot\cdot\cdot\cdot\cdot \\
a_{n1}\ a_{n2}\ \qquad\quad a_{nm}
\end{bmatrix}
\quad
X = \begin{bmatrix} x_1 \\ x_2 \\ \cdot\cdot \\ \cdot\cdot \\ x_m \end{bmatrix}
\quad
Y = \begin{bmatrix} y_1 \\ y_2 \\ \cdot\cdot \\ \cdot\cdot \\ y_n \end{bmatrix}
$$

By premultiplying eqn. 2.43 by A^T (the transpose of matrix A), we obtain

$$
\underbrace{A^\mathrm{T}}_{m \times n}\ \underbrace{A}_{n \times m}\ \underbrace{X}_{m \times 1}\ =\ \underbrace{A^\mathrm{T}}_{m \times n}\ \underbrace{Y}_{n \times 1}
\tag{2.44}
$$

In its expanded form, eqn. 2.44 is of the form

$$
\begin{bmatrix}
\sum a_{i1}a_{i1} & \sum a_{i1}a_{i2} & \cdots & \sum a_{i1}a_{im} \\
\sum a_{i2}a_{i1} & \sum a_{i2}a_{i2} & \cdots & \sum a_{i2}a_{im} \\
\cdot\cdot\cdot\cdot\cdot\cdot & \cdot\cdot\cdot\cdot\cdot\cdot & & \cdot\cdot\cdot\cdot\cdot\cdot \\
\sum a_{im}a_{i1} & \sum a_{im}a_{i2} & \cdots & \sum a_{im}a_{im}
\end{bmatrix}
\begin{bmatrix} x_1 \\ x_2 \\ \cdot\cdot\cdot \\ x_m \end{bmatrix}
=
\begin{bmatrix} \sum a_{i1}y_i \\ \sum a_{i2}y_i \\ \cdot\cdot\cdot\cdot\cdot\cdot \\ \sum a_{im}y_i \end{bmatrix}
$$

$$\tag{2.45}$$

It should be noted that the limits of all the summations in eqn. 2.45 are from $i = 1$ to n. Eqn. 2.45 in effect defines the normal equations obtained by using the least-squares method. It can also be seen from eqn. 2.44 that the matrix $A^\mathrm{T}A$ is a square matrix having a dimension of $m \times m$. Therefore, by premultiplying eqn. 2.44 by $[A^\mathrm{T}A]^{-1}$, we obtain the vector of unknowns as given by eqn. 2.46.

$$
\underbrace{X}_{m \times 1}\ =\ \underbrace{[A^\mathrm{T}\ A]^{-1}}_{m \times m}\ \underbrace{A}_{m \times n}\ \underbrace{Y}_{n \times 1}
\tag{2.46}
$$

The matrix $[A^\mathrm{T}A]^{-1}A$ is known as the left pseudoinverse of A and, since eqn. 2.46 is the solution of the least-square error eqn. 2.45, it is therefore clear that the pseudoinverse approach provides the least-squares solution.

2.5.2 Smoothing [3]

Data obtained from physical measurement often contain errors that can be conveniently defined as the difference between the correct and measured values. The correct value means the true, theoretical or average value. In practice, it is usually desirable to eliminate as much error as possible from the measured data, and this can be done by fitting the data to a formula derived using the least-squares method. This process is often known as smoothing.

Example
Let us assume that it is required to find the smoothed value for a measured three-point data set using linear equations. Let the measured data be denoted by (x_1, y_1), (x_2, y_2) and (x_3, y_3). The process of finding the smoothed values of the given data to fit a linear equation basically consists of the following steps:

(a) The data are fitted, using the least squares method, to the linear function of eqn. 2.47.

$$y = a_0 + a_1 x \qquad (2.47)$$

This of course involves the determination of the constants a_0 and a_1 to give the best fit.

(b) The smoothed value Y_k can then be determined at the corresponding x_k by using eqn. 2.48.

$$Y_k = a_0 + a_1 x_k \qquad (2.48)$$

To simplify the analysis let us:
(i) Assume that the data are taken at equal spaces in the x variable, such that

$$x_k = kh \qquad (2.49)$$

where h is the interval between values.
(ii) Renumber the measured data in such a way as to extend them symmetrically from negative to positive around a defined middle point. Therefore the data set becomes (x_{-1}, y_{-1}), (x_0, y_0), (x_1, y_1).

From previous work (see eqn. 2.41), the measured data set can be fitted to eqn. 2.47, using the least-squares method, to obtain the following normal equations:

$$\sum_{i=1}^{N} y_i = N a_0 + \sum_{i=1}^{N} a_1 x_i \qquad (2.50)$$

$$\sum_{i=1}^{N} x_i y_i = \sum_{i=1}^{N} a_0 x_i + \sum_{i=1}^{N} a_1 x_i^2 \qquad (2.51)$$

where N is the number of measured data points (here $N=3$).
　By substituting eqn. 2.49 into eqn. 2.50 and 2.51, and changing the limits as indicated above, the normal equations become:

$$\sum_{k=-1}^{1} y_k = N a_0 + a_1 \sum_{k=-1}^{1} (kh) \qquad (2.52)$$

$$\sum_{k=-1}^{1} (kh)y_k = a_0 \sum_{k=-1}^{1} (kh) + a_1 \sum_{k=-1}^{1} (kh)^2 \qquad (2.53)$$

From eqn. 2.52 we obtain

$$y_{-1} + y_0 + y_1 = 3a_0 + a_1(-kh + 0 + kh)$$

which gives

$$a_0 = (y_{-1} + y_0 + y_1)/3 \qquad (2.54)$$

From eqn. 2.53 we likewise obtain

$$h(-y_{-1} + 0 + y_1) = a_0[-h + 0 + h] + a_1[h^2 + 0 + h^2]$$

which likewise gives

$$a_1 = (-y_{-1} + y_1)/2h \qquad (2.55)$$

The smoothed values Y_k corresponding to $x_k = kh(k = -1, 0, 1)$ can be found by substituting eqn. 2.54 and eqn. 2.55 into eqn. 2.48. Thus

$$Y_{-1} = (y_{-1} + y_0 + y_1)/3 + (y_{-1} - y_1)/2$$

Therefore

$$Y_{-1} = (5y_{-1} + 2y_0 - y_1)/6 \qquad (2.56)$$

Similarly the smoothed values Y_0 and Y_1 can be found as

$$Y_0 = (y_{-1} + y_0 + y_1)/3 \qquad (2.57)$$

$$Y_1 = (-y_{-1} + 2y_0 + 5y_1)/6 \qquad (2.58)$$

2.6 Fourier analysis [10, 11]

2.6.1 The Fourier series

Trigonometric form
Any periodic function $f(t)$ can normally be represented by a Fourier series of discrete harmonics

$$f(t) = \frac{a_0}{2} + \sum_{n=1}^{\infty} a_n \cos n\omega_0 t + \sum_{n=1}^{\infty} b_n \sin n\omega_0 t \qquad (2.59)$$

where ω_0 is the angular fundamental frequency $= 2\pi f_0 = 2\pi/T$
   T is the time period of the fundamental component
   $n\omega_0$ is the nth harmonic angular frequency
   t_1 is arbitrary

Given the known function $f(t)$ (which in most practical situations is a function that varies with time), the coefficients $a_0, a_1, b_1, \ldots a_n, b_n$ can be determined from expressions of the form of eqns 2.60–2.62 [10, 11].

$$a_0 = \frac{2}{T} \int_{t_1}^{t_1+T} f(t) \, dt \qquad (2.60)$$

$$a_n = \frac{2}{T} \int_{t_1}^{t_1+T} f(t) \cos n\omega_0 t \, dt \qquad (2.61)$$

$$b_n = \frac{2}{T} \int_{t_1}^{t_1+T} f(t) \sin n\omega_0 t \, dt \qquad (2.62)$$

Alternatively, by combining corresponding sine and cosine terms of the same frequency, eqn. 2.59 can be written as

$$f(t) = \sum_{n=0}^{\infty} A_n \cos(n\omega_0 t + \theta_n) \qquad (2.63)$$

where

$$A_0 = \frac{a_0}{2}, \; \theta_0 = 0 \qquad (2.64)$$

and

$$A_n = \sqrt{a_n^2 + b_n^2}, \; \theta_n = \tan^{-1} \frac{b_n}{a_n} \; (n = 1, 2, \ldots) \qquad (2.65)$$

Complex form
In some applications it is more convenient to use the complex form of Fourier series given in eqn. 2.66. It will be apparent that the latter equation is directly equivalent to the basic Fourier series expression given in eqn. 2.59.

$$f(t) = \sum_{n=-\infty}^{\infty} \bar{F}_n \, e^{jn\omega_0 t} \qquad (2.66)$$

where

$$\bar{F}_n = (a_n - jb_n)/2, \; n = \pm 1, \pm 2 \ldots \pm \infty$$
$$= a_0/2, \; n = 0$$

By substituting eqns. 2.60–2.62 into the above equation, $\bar{F}_n$ reduces to

$$\bar{F}_n = \frac{1}{T} \int_{t_1}^{t_1+T} f(t) \, e^{-jn\omega_0 t} dt, \; n = \pm 1, \pm 2 \ldots \pm \infty \qquad (2.67)$$

$$= \frac{1}{T} \int_{t_1}^{t_1+T} f(t) dt, \; n = 0$$

The spectrum of periodic signals
It will be apparent from the previous section that any periodic function $f(t)$ can be represented in one of two ways:

(i) a time-domain representation, where the time history of the function behaviour is described by using the equation which defines the function $f(t)$

(ii) a frequency-domain representation, where the waveform is described by the magnitude and phase of a number of sinusoidal components of frequencies $n\omega_0 (n = 1, 2, \ldots)$ that make up the signal. This is done by specifying all the Fourier coefficients, i.e. any one of the three sets (a_n, b_n), (A_n, θ_n) or $\bar{F}_n$.

Example

Figure 2.3 shows a periodic function $f(t)$ with a period of T, which is given by

$$f(t) = \begin{cases} A & \left| mT - \dfrac{\tau}{2} \right| \leq t \leq \left| mT + \dfrac{\tau}{2} \right| & m = 0, \pm 1, \pm 2, \ldots \pm \infty \\ 0 & \text{elsewhere} \end{cases} \qquad (2.68)$$

Find its representation in the frequency domain.

Solution

Representation of the function $f(t)$ in the frequency domain requires the determination of its Fourier coefficients:

From eqn. 2.67

$$\bar{F}_n = \frac{1}{T} \int_{t_1}^{t_1+T} f(t) \ e^{-jn\omega_0 t} dt, \ n = \pm 1, \pm 2, \ldots \pm \infty$$

$$= \frac{1}{T} \int_{t_1}^{t_1+T} f(t) dt, \ n = 0$$

Using the definition of $f(t)$ given in eqn. 2.68 and taking the limits of integration from $-\tau/2$ to $\tau/2$, we obtain

$$\bar{F}_n = \frac{1}{T} \int_{-\tau/2}^{\tau/2} A \ e^{-jn\omega_0 t} \ dt, \ n = \pm 1, \pm 2, \ldots \pm \infty$$

$$= \frac{A}{T} \left[\frac{-jn\omega_0 t}{-jn\omega_0} \right]_{-\tau/2}^{\tau/2}$$

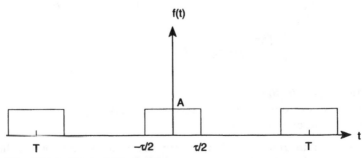

Figure 2.3   Illustrative periodic function

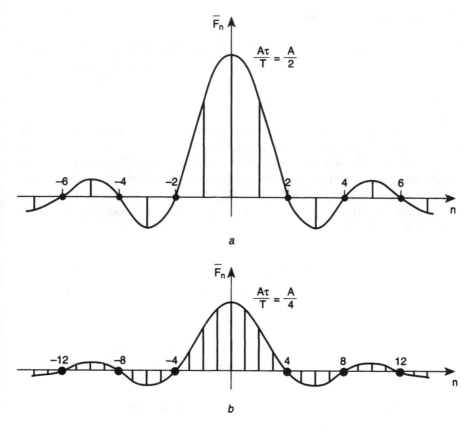

Figure 2.4   *Fourier coefficients F_n ($n = 0, 1, \ldots$) versus n for the periodic function shown in Fig. 2.3 for (a) $\tau/T = 1/2$, and (b) $\tau/T = 1/4$*

$$= \frac{A\tau}{T} \operatorname{sinc}(n\omega_0\tau/2) = \frac{A\tau}{T} \operatorname{sinc}(n\pi\tau/T)$$

$$= \frac{A\tau}{T}, \quad n = 0 \tag{2.69}$$

where

$$\operatorname{sinc}(x) = \frac{\sin x}{x}$$

The value of $\bar{F}_n$ ($n = 0, \pm 1, \pm 2, \ldots$) defines the Fourier coefficients corresponding to discrete angular frequencies ω_n, $n = 0, 1, 2, \ldots$. By knowing the ratio τ/T, it is possible to determine F_n and Figure 2.4(a) and (b) shows $\bar{F}_n$ versus ω_n for ratios $\tau/T = 1/2$ and $\tau/T = 1/4$ respectively. These figures also represent the function $f(t)$ in the frequency domain.

2.6.2 The Fourier transform

We have seen in previous Sections how any periodic function can be treated using the Fourier series. However, in practice, non-periodic functions often arise. This is particularly so in protection, where transient waveforms have to be processed. Therefore the question arises as to how non-periodic functions can be treated.

The first step in answering this question is to note that a non-periodic function can be regarded as one having a periodic time interval T equal to infinity $(T \to \infty)$. It follows that $f(t) = \lim_{T \to \infty} f_p(t)$, where $f_p(t)$ is a replica of the single function that is repeated after an assumed infinite time. If we assume that the Fourier coefficients $\bar{F}_n$ of the function $f_p(t)$ are known, then $f_p(t)$ itself can be found from substituting eqn. 2.67 into eqn. 2.66. Therefore by assuming $t_1 = -T/2$, we obtain

$$f_p(t) = \sum_{n=-\infty}^{\infty} \left(\frac{1}{T} \int_{-T/2}^{T/2} f_p(t) \, e^{-jn\omega_0 t} \, dt \right) e^{jn\omega_0 t}$$

or

$$f_p(t) = \frac{1}{2\pi} \sum_{n=-\infty}^{\infty} \left(\int_{-T/2}^{T/2} f_p(t) \, e^{-jn\omega_0 t} \, dt \right) e^{jn\omega_0 t} \cdot \frac{2\pi}{T} \qquad (2.70)$$

Now, on allowing $T \to \infty$, the following observations apply:

(i) the fundamental frequency $\omega_0 = \dfrac{2\pi}{T} = d\omega \to 0$ and $n\omega_0 = \omega$

(ii) an infinite number of harmonics spaced by $d\omega \to 0$ are required to represent the function $f(t)$. In essence, the time function is represented by a continuous spectrum of frequencies all separated by an infinitesimally small frequency space $d\omega$.

It follows from the foregoing and eqn. 2.70 that

$$f(t) = \lim_{T \to \infty} f_p(t)$$

$$= \lim_{T \to \infty} \frac{1}{2\pi} \sum_{n=-\infty}^{\infty} \left(\int_{-T/2}^{T/2} f_p(t) \, e^{-jn\omega_0 t} dt \right)^{jn\omega_0 t} \frac{2\pi}{T} \qquad (2.71)$$

or

$$f(t) = \frac{1}{2\pi} \int_{-\infty}^{\infty} \left(\int_{-\infty}^{\infty} f(t) \, e^{-j\omega t} dt \right) e^{jn\omega t} \, d\omega \qquad (2.72)$$

The term between the brackets in eqn. 2.72 is seen to be a function of frequency and will be denoted by $F(\omega)$, i.e.

$$F(\omega) = \int_{-\infty}^{\infty} f(t) \, e^{-j\omega t} \, dt \tag{2.73}$$

The frequency function $F(\omega)$ is called the Fourier transform of $f(t)$. In terms of the Fourier transform, the function $f(t)$ can thus be written as

$$f(t) = \frac{1}{2\pi} \int_{-\infty}^{\infty} F(\omega) \, e^{j\omega t} \, d\omega \tag{2.74}$$

The time domain function $f(t)$ and the corresponding frequency domain function $F(\omega)$ are commonly known as transform pairs, and any non-periodic function of time can be transformed to a continuous frequency spectrum $F(\omega)$ via the Fourier transform integral of eqn. 2.73. The inverse Fourier transform likewise inverts the process by transforming a continuous frequency spectrum into the corresponding time variation.

Example
Find the Fourier transform of the rectangular pulse shown in Figure 2.5(a).

Solution
The pulse shown in Figure 2.5(a) can be expressed mathematically as

$$P_\tau(t) = \begin{cases} 1 & -\tau/2 \leq t \leq \tau/2 \\ 0 & \text{elsewhere} \end{cases}$$

The Fourier transform of $P_\tau(t)$ is obtained from eqn. 2.73 as

$$P_\tau(\omega) = \int_{-\infty}^{\infty} P_\tau(t)^{-j\omega t} \, dt$$

$$= \int_{-\tau/2}^{\tau/2} e^{-j\omega t} \, dt = \left[\frac{e^{-j\omega t}}{-j\omega} \right]_{-\tau/2}^{\tau/2}$$

$$= \frac{e^{+j\omega\tau/2} - e^{-j\omega\tau/2}}{j\omega}$$

which can be converted to the form

$$P_\tau(\omega) = \tau \frac{\sin(\omega\tau/2)}{\omega\tau/2} = \tau \operatorname{sinc} \frac{\omega\tau}{2} \tag{2.75}$$

Figure 2.5(b) shows how $P_\tau(\omega)$ varies with angular frequency. It is interesting to note that the shorter the pulse, the more quickly the Fourier transform falls to zero, i.e. as the pulse duration τ is reduced, the magnitudes of spectral components also reduce and the distances between consecutive crossings of the ω-axis become larger. This is as expected since a reduction in the pulse width is accompanied by a reduction in signal energy.

2.7 Walsh function analysis [12, 13, 14]

The Fourier series approach enables a periodic function to be represented by a summation of discrete sinusoidal components. An alternative approach enables a periodic function to be represented by a series of orthogonal square waves known as Walsh functions. The Walsh function is consequently useful where signals take the form of logic pulses, which are fundamentally rectangular in

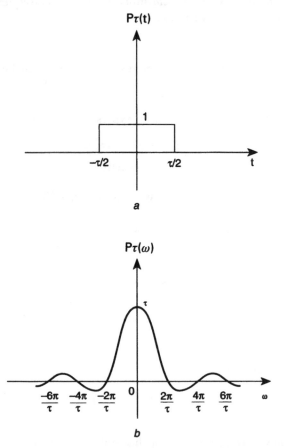

Figure 2.5   (a) Non-periodic (transient) pulse and (b) corresponding Fourier transform

nature. Conversely, it is relatively easy to generate square (or rectangular) pulse trains, and it is consequently relatively easy to construct or analyse a given waveform using the Walsh technique.

2.7.1 Definition of Walsh functions

The Walsh function of order k (written as $\text{Wal}(k, t/T)$) is a time function having a magnitude of either $+1$ or -1 that crosses the time axis k times per unit interval T (i.e. the number of times it changes its sign per unit interval T is equal to k).

Figure 2.6 shows the first eight Walsh functions of integral index k for $0 \leqslant t \leqslant T$. The number of sign changes per unit time is called the sequency

$$S = k/T$$

2.7.2 Some fundamental properties of Walsh functions

(a) Index and timescale
Doubling the index of a Walsh function is equivalent to compressing the timescale by a factor of half, or

$$\text{Wal}(2k, t/T) = \text{Wal}(k, 2t/T) \tag{2.76}$$

(b) Multiplication
When two Walsh functions are multiplied together, another Walsh function is produced such that

$$\text{Wal}(h, t/T) \cdot \text{Wal}(k, t/T) = \text{Wal}(h \otimes k, t) \tag{2.77}$$

where $h \otimes k$ denotes a number whose binary numeral has 0s in those positions where the binary numerals for h and k are alike and has 1s where they are different. For example

$$\text{Wal}(3, t/T) \, \text{Wal}(5, t/T) = \text{Wal}(6, t/T)$$

Because 3 (in decimal) $= 011$ (in binary), and 5 (in decimal) $= 101$ (in binary), $3 \otimes 5 = 110$ (binary) $= 6$ (decimal).

(c) Symmetry
With reference to Figure 2.6, it can be concluded that

$$\text{Wal}\left(k, \frac{T-t}{T}\right) = (-1)^k \, \text{Wal}(k, t/T) \tag{2.78}$$

Thus $\text{Wal}(k, t/T)$ has even or odd symmetry about $t = T/2$ depending on whether k is even or odd.

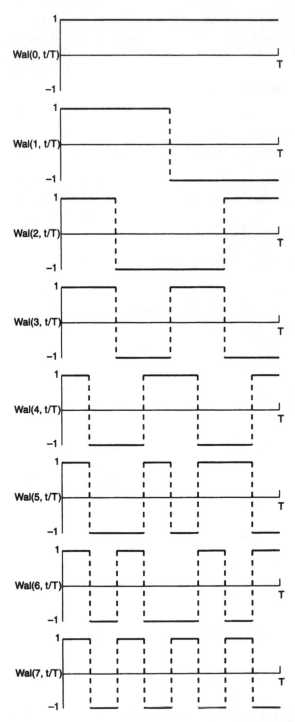

Figure 2.6   The first eight Walsh functions of integral index k for $0 \leqslant t \leqslant T$

2.7.3 *Discrete representation of Walsh functions*

In digital applications, it is useful to represent Walsh functions in discrete form. Let N be the number of samples per unit interval, i.e., the length of the discrete Walsh function is N. If $N = 2^P$, then the discrete Walsh function $\text{Wal}(k, j/T)$ can be defined as

$$\text{Wal}(k, j/T) = \prod_{r=0}^{P} (-1)^{(k_{p-r} + k_{p-r-1})j_r} \tag{2.79}$$

with

$$j_r = 0, 1, 2, \ldots\ldots, N-1$$

$$k = 0, 1, 2, \ldots\ldots, N-1$$

where $\prod$ is the factorisation symbol in which the parameters that follow are multiplied together. $k_r, j_r,$ are binary bits of k, j or

$$k = \sum_{r=0}^{p-1} k_r \cdot 2^r \quad \text{and} \quad j = \sum_{r=0}^{p-1} j_r \cdot 2^r \tag{2.80}$$

This definition can be best illustrated by the following example.

Example
Find the fourth element of the Walsh function having the order of 5 and length $N = 8$, i.e., $\text{Wal}(5, 4)$.

Solution

$$N = 8 = 2^3$$

therefore

$$P = 3$$

$$k = 5 \text{ (in decimal)}$$

$$= 101 \text{ (in binary)} = 1 \times 2^2 + 0 \times 2^1 + 1 \times 2^0$$

$$k_0 = 1, k_1 = 0, k_2 = 1$$

Similarly

$$j = 4 \text{ (in decimal)}$$

$$= 100 \text{ (in binary)}$$

$$= 1 \times 2^2 + 0 \times 2^1 + 0 \times 2^0$$

hence

$$j_0 = 0, j_1 = 0, j_2 = 1$$

Now Wal(5, 4) can be found from eqn. 2.79

$$\text{Wal}(5, 4) = \prod_{r=0}^{3} (-1)^{(k_{3-r} + k_{3-r-1})j_r}$$

$$= (-1)^{(k_3 + k_2)j_0} \cdot (-1)^{(k_2 + k_1)j_1} \cdot (-1)^{(k_1 + k_0)j_2}$$

$$= (-1)^0 \cdot (-1)^0 \cdot (-1)^1 \quad (\text{because } j_0 = j_1 = 0)$$

$$= -1$$

Using this procedure it is possible to construct an 8×8 matrix for an 8-length eight function:

$$\text{Wal}(k, j) = \begin{array}{c} \\ \\ \end{array} \begin{array}{cccccccc} j & 0 & 1 & 2 & 3 & 4 & 5 & 6 & 7 \end{array}$$

$$\text{Wal}(k, j) = \begin{bmatrix} 1 & 1 & 1 & 1 & 1 & 1 & 1 & 1 \\ 1 & 1 & 1 & 1 & -1 & -1 & -1 & -1 \\ 1 & 1 & -1 & -1 & -1 & -1 & 1 & 1 \\ 1 & 1 & -1 & -1 & 1 & 1 & -1 & -1 \\ 1 & -1 & -1 & 1 & 1 & -1 & -1 & 1 \\ 1 & -1 & -1 & 1 & -1 & 1 & 1 & -1 \\ 1 & -1 & 1 & -1 & -1 & 1 & -1 & 1 \\ 1 & -1 & 1 & -1 & 1 & -1 & 1 & -1 \end{bmatrix} \begin{array}{c} k \\ 0 \\ 1 \\ 2 \\ 3 \\ 4 \\ 5 \\ 6 \\ 7 \end{array}$$

2.7.4 The Walsh series

In terms of Walsh functions, a periodic function say $f(t)$ can be expanded over the interval $(t_1, t_1 + T)$ as follows:

$$f(t) = \sum_{k=0}^{\infty} W_k \, \text{Wal}(k, t/T) \qquad (2.81)$$

where the Walsh coefficient W_k is given by

$$W_k = \frac{1}{T} \int_{t_1}^{t_1 + T} f(t) \, \text{Wal}(k, t/T) \, dt \qquad (2.82)$$

where W_k is the kth coefficient of the Walsh series and Wal(k, t/T) is the kth order Walsh function. This technique mirrors the previously described method of constructing a periodic waveform by a number of sinusoidal components using the Fourier series technique.

2.8 Relationship between Fourier and Walsh coefficients

In this Section we establish the interrelationship between Fourier and Walsh coefficients, i.e. the transformation matrix which transforms a Walsh coefficient vector **W** into its equivalent Fourier coefficient vector **F**.

Let us first write a Fourier series defined by eqn. 2.59 in terms of the Fourier coefficients $F_i (i = 0, 1, 2, \ldots)$, such that

$$f(t) = F_0 + \sqrt{2} F_1 \sin \omega_0 t + \sqrt{2} F_2 \cos \omega_0 t + \sqrt{2} F_3 \sin 2\omega_0 t + \sqrt{2} F_4$$

$$\cos 2\omega_0 t + \cdots \cdots \tag{2.83}$$

It is easily verified that

$$\left. \begin{array}{l} F_0 = \dfrac{a_0}{2} = \dfrac{1}{T} \displaystyle\int_0^T f(t)\ dt \\[4mm] F_{2n-1} = \dfrac{b_n}{\sqrt{2}} = \dfrac{\sqrt{2}}{T} \displaystyle\int_0^T f(t)\ \sin n\omega_0 t\ dt,\ n = 1, 2, \ldots \\[4mm] F_{2n} = \dfrac{a_n}{\sqrt{2}} = \dfrac{\sqrt{2}}{T} \displaystyle\int_0^T f(t)\ \cos n\omega_0 t\ dt,\ n = 1, 2, \ldots \end{array} \right\} \tag{2.84}$$

If the same function $f(t)$ is expanded using Walsh series, we have:

$$f(t) = \sum_{k=0}^{\infty} W_k\ \mathrm{Wal}(k, t/T) \tag{2.85}$$

Now it is possible to calculate Walsh coefficients (W_k) by substituting eqn. 2.83 into eqn. 2.82 such that

$$W_0 = \frac{1}{T} \int_{t_1}^{t_1+T} [F_0 + \sqrt{2} F_1 \sin \omega_0 t + \sqrt{2} F_2 \cos \omega_0 t + \sqrt{2} F_3 \sin 2\omega_0 t$$

$$+ \ldots] \mathrm{Wal}(0, t/T) dt$$

$$W_1 = \frac{1}{T} \int_{t_1}^{t_1+T} [F_0 + \sqrt{2} F_1 \sin \omega_0 t + \sqrt{2} F_2 \cos \omega_0 t + \sqrt{2} F_3 \sin 2\omega_0 t$$

$$+ \ldots] \mathrm{Wal}(1, t/T) dt$$

$$\vdots$$

$$W_k = \frac{1}{T} \int_{t_1}^{t_1+T} [F_0 + \sqrt{2} F_1 \sin \omega_0 t + \sqrt{2} F_2 \cos \omega_0 t + \sqrt{2} F_3 \sin 2\omega_0 t$$

$$+ \ldots] \mathrm{Wal}(k, t/T) dt$$

Using matrix notation, these equations can be written as

$$
\begin{bmatrix} W_0 \\ W_1 \\ W_2 \\ \cdot\cdot \\ W_k \\ \cdot\cdot \\ \cdot\cdot \end{bmatrix} = \begin{bmatrix} A_{00} & A_{01} & A_{02} & \cdots & A_{0k}, & \cdots \\ A_{10} & A_{11} & A_{12} & \cdots & A_{1k}, & \cdots \\ A_{20} & A_{21} & A_{22} & \cdots & A_{2k}, & \cdots \\ \cdots & \cdots & \cdots & \cdots & \cdots & \cdots \\ A_{k0} & A_{k1} & A_{k2} & \cdots & A_{kk}, & \cdots \\ \cdots & \cdots & \cdots & \cdots & \cdots & \cdots \\ \cdots & \cdots & \cdots & \cdots & \cdots & \cdots \end{bmatrix} \begin{bmatrix} F_0 \\ F_1 \\ F_2 \\ \cdot\cdot \\ F_k \\ \cdot\cdot \\ \cdot\cdot \end{bmatrix}
\qquad (2.86)
$$

or

$$
W = AF
$$

where

$$
\left.
\begin{aligned}
A_{k0} &= \frac{1}{T} \int_{t_1}^{t_1+T} \mathrm{Wal}(k, t/T)dt \\[2mm]
A_{k,\,2m-1} &= \frac{\sqrt{2}}{T} \int_{t_1}^{t_1+T} \sin(m\omega_0)t \cdot \mathrm{Wal}(k, t/T)dt \\[2mm]
A_{k,\,2m} &= \frac{\sqrt{2}}{T} \int_{t_1}^{t_1+T} \cos(m\omega_0)t \cdot \mathrm{Wal}(k, t/T)dt, \quad m = 1, 2, 3, \ldots
\end{aligned}
\right\}
\qquad (2.87)
$$

Eqn. 2.87 can be simplified by taking the lower time limit $t_1 = 0$, and normalising the time with respect to the period T, such that

$$
t' = \frac{t}{T}, \quad dt = T dt'
\qquad (2.88)
$$

When eqn. 2.88 is substituted into eqn. 2.87 (replacing the subscript $2m$ by K and simplifying) we obtain:

$$
A_{k0} = \int_0^1 \mathrm{Wal}(k, t')dt'
$$

$$
A_{k,\,K-1} = \sqrt{2} \int_0^1 \sin(K\pi)t' \, \mathrm{Wal}(k, t')dt'
\qquad (2.89)
$$

$$
A_{k,\,K} = \sqrt{2} \int_0^1 \cos(K\pi)t' \, \mathrm{Wal}(k, t')dt'
$$

By direct substitution of the Walsh functions $\mathrm{Wal}(k, t')$, shown in Figure 2.6, it is possible to calculate the elements of the matrix A having any given size. For example the elements of a 12×16 A-matrix have been evaluated using this approach to obtain the matrix eqn. 2.90 (see page 37).

As indicated in eqn. 2.86, the relationship between Fourier and Walsh coefficients is given in matrix form as

$$
W = AF
\qquad (2.91)
$$

$$
A =
\begin{array}{c|ccccccccccccccccc|c}
n\backslash k & 0 & 1 & 2 & 3 & 4 & 5 & 6 & 7 & 8 & 9 & 10 & 11 & 12 & 13 & 14 & 15 & 16 & \\
\hline
0 & 0 & 1 & 0 & 0 & 0 & 0 & 0 & 0 & 0 & 0 & 0 & 0 & 0 & 0 & 0 & 0 & 0 & \\
1 & 1 & 0 & 0 & 0 & 0 & 0 & 0 & 0 & 0 & 0 & 0 & 0 & 0 & 0 & 0 & 0 & 0 & \\
2 & 0 & 0.900 & 0.900 & 0 & 0 & 0.300 & -0.300 & 0 & 0 & 0.180 & 0.180 & 0 & 0 & 0.129 & -0.129 & 0 & 0 & \\
3 & 0 & 0 & 0 & 0.900 & 0 & 0 & 0 & 0 & 0 & 0 & 0 & 0.300 & -0.300 & 0 & 0 & 0 & 0 & \\
4 & 0 & 0 & 0 & 0 & 0.900 & 0 & 0 & 0 & 0 & 0 & 0 & 0 & 0 & 0 & 0 & 0 & 0 & \\
5 & 0 & -0.373 & 0.373 & 0 & 0 & 0.724 & 0.724 & 0 & 0 & 0.435 & -0.435 & 0 & 0 & -0.053 & -0.053 & 0 & 0 & \\
6 & 0 & 0 & 0 & 0 & 0 & 0 & 0 & 0 & 0 & 0 & 0 & 0 & 0 & 0 & 0 & 0 & 0 & \\
7 & 0 & 0 & 0 & 0 & 0 & 0 & 0 & 0.900 & 0 & 0 & 0 & 0 & 0 & 0 & 0 & 0 & 0 & \\
8 & 0 & 0 & 0 & 0 & 0 & 0 & 0 & 0 & 0.900 & 0 & 0 & 0 & 0 & 0 & 0 & 0 & 0 & \\
9 & 0 & -0.074 & -0.074 & 0 & 0 & -0.484 & 0.484 & 0 & 0 & 0.650 & 0.650 & 0 & 0 & 0.268 & -0.268 & 0 & 0 & \\
10 & 0 & 0 & 0 & 0 & 0 & 0 & 0 & 0 & 0 & 0 & 0 & 0 & 0 & 0 & 0 & 0 & 0 & \\
11 & 0 & 0 & 0 & 0 & 0 & 0 & 0 & 0 & 0 & 0 & 0 & 0 & 0 & 0 & 0 & 0 & 0 & \\
12 & 0 & 0 & 0 & -0.373 & 0 & 0 & 0 & 0 & 0 & 0 & 0 & 0.724 & 0 & 0 & 0 & 0 & 0 & \\
13 & 0 & 0 & 0 & 0 & 0.373 & 0.373 & 0 & 0 & 0 & 0 & 0 & 0 & 0.724 & 0 & 0 & 0 & 0 & \\
\hdashline
\end{array}
\qquad (2.90)
$$

where W is the Walsh coefficient vector, F is the Fourier coefficient vector, A is the matrix that transforms F-vectors into W-vectors and whose elements are defined by eqn. 2.90.

Fourier coefficients can likewise be determined from Walsh coefficients by using the fact that $A^{-1} = A^T$. Thus the inverse relationship is

$$F = A^T W \qquad (2.92)$$

where A^T is the transpose of matrix A.

2.9 References

1 KREYSZIG, E.: 'Advanced engineering mathematics' (5th Edition, John Wiley & Sons, 1983)
2 WYLIE, C.R.: 'Advanced engineering mathematics' (4th Edition, McGraw-Hill, 1975)
3 LaFARA, R.L.: 'Computer methods for science and engineering' (Intertext Books, 1973)
4 HILDEBRAND, F.B.: 'Introduction to numerical analysis' (2nd Edition, McGraw-Hill, 1974)
5 GERALD, C.E.: 'Applied numerical analysis' (4th Edition, Addison-Wesley, 1989)
6 APOSTOL, T.M.: 'Calculus—introduction with vectors and analytic geometry', Vol 1 (3rd Edition, Blaisdell Publishing Company, 1962)
7 RALSTON, A.: 'A first course in numerical analysis' (2nd Edition, McGraw-Hill, 1978)
8 KAPLAN, W. 'Advanced mathematics for engineers' (Addison-Wesley, 1981)
9 STANG, G.: 'Linear algebra and its applications' (Academic Press, New York, 1976)
10 MAYHAN, R.J.: 'Discrete-time and continuous-time linear systems' (Addison-Wesley, 1984)
11 McGILLEM, C.D., and COOPER, G.R.: 'Continuous and discrete signal and system analysis' (2nd Edition, Holt, Rinehart & Winston, 1984)
12 BLACKMAN, N.M.: 'Sinusoids versus Walsh functions' *Proc. IEEE*, 1974, **62**, pp. 346–354
13 SCHREIBER, H.H.: 'Bandwidth requirements for Walsh functions' *IEEE Trans.*, 1970, **IT-16**, pp. 491–493
14 KENNET, B.L.N.: 'Note on the finite Walsh transform' *IEEE Trans.* 1970, **IT-16**, pp. 489–491

Chapter 3
Basic elements of digital protection

3.1 Introduction

Operating voltages and currents flowing through a power system are usually at kilovolt and kiloampere levels. However, for digital processing, it is necessary to reduce the primary measurands to manageable levels. Therefore, the analogue signals are converted to digital form, thereby allowing subsequent digital processing to be performed to determine the circuit state.

In this Chapter the basic principles underlying the conversion of analogue signals into equivalent digital forms will be explained. We shall also explain the essentially common features of various digital relaying schemes, other detailed aspects being discussed in later chapters.

3.2 Basic components of a digital relay

Any digital relay can be thought of as comprising three fundamental subsystems (Figure 3.1):

(i)   a signal conditioning subsystem
(ii) a conversion subsystem
(iii) a digital processing relay subsystem.

The first two subsystems are generally common to all digital protective schemes, while the third varies according to the application of a particular scheme. Each of the three subsystems is built up of a number of components and circuits, as discussed in detail in the following sections.

3.3 Signal conditioning subsystem [1–5]

3.3.1 Transducers

Primary power system currents and voltages are usually relatively high. Before it is possible to bring these signals to protective relays, they must therefore be reduced to much lower levels. Conventionally, currents are reduced either to 5 A or 1 A and voltages are reduced to 110 V or 120 V. This is normally achieved by using primary current and voltage transducers (CTs and VTs). In digital relays, however, current and voltage magnitudes are both further

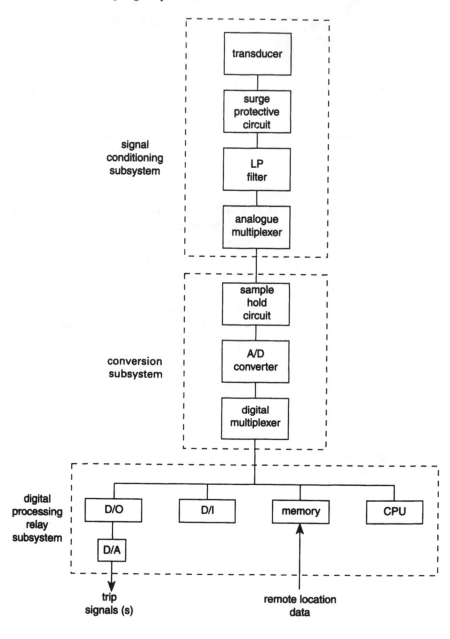

LP = low pass
A/D = analogue-to-digital
D/A = digital-to-analogue
CPU = central processor unit
D/I = data input
D/O = data output

Figure 3.1   Basic components of a digital relay

reduced using auxiliary transducers and/or mimic impedances within the relays to suit the requirements of the components used.

Ideally the current transformer would reproduce a perfectly scaled down version of the primary signal on its secondary side. Practical transformers reproduce the secondary current with some error, because these devices incorporate 'non-ideal' elements. The worst condition occurs when the ironcore saturates during faults. The degree of signal distortion and the time after a fault at which it occurs is heavily dependent on the total burden connected to the primary current transducers. However, in most modern practical applications, the overall burden is such that the current signal distortion is small during the measuring period. In cases where this is not so, it is desirable to establish the effect on performance and, if necessary, incorporate means within the relay software to ensure that integrity of measurement is maintained. Some work on compensating for current transducer saturation is to be found in the literature, although, in most applications, this is unnecessary. In what follows, the effect of current transducer saturation will, for brevity, be assumed negligible.

Electromagnetic VTs generally produce a very high-fidelity voltage signal, though in fact these are rarely used at system voltages above typically 100 kV. At higher voltages, the use of capacitor voltage transformers (CVTs) is commonplace. Unfortunately, the transient response of CVTs varies widely according to the type of transducer involved and the nature of the total connected burden. In high-speed relaying applications in particular, account needs to be taken of the rather poor fidelity of the voltage signals that emanate from CVTs, there being many examples in the literature by which the digital algorithm compensates for such effects. A detailed consideration of such techniques is not necessary in a text of this type.

3.3.2 Surge protection circuits

The current and voltage from the secondaries of the CTs and VTs is connected to surge protective circuits, which typically consist of capacitors and isolating transformers (Figure 3.2). Zener diodes are also commonly used to protect electronic circuits against surges, though their placement depends on the exact physical circuit arrangement used. In practice, it is common to convert the secondary current measurands into low-level voltage signals by means of a suitably connected burden and/or current-to-voltage amplifier arrangement. The latter normally use careful screening techniques and are often accommodated inside separately screened self-contained modules separated from the digital signal processing hardware.

3.3.3 Analogue filtering

It is normally necessary to perform analogue filtering of the signals received from the CTs and VTs. In practice, the amount of filtering depends on the data requirements of a particular digital relay. Such filtering is usually performed using low-pass filters to remove unwanted high frequencies before sampling. In fact, as will be shown in Section 3.4, the sampling theorem requires that

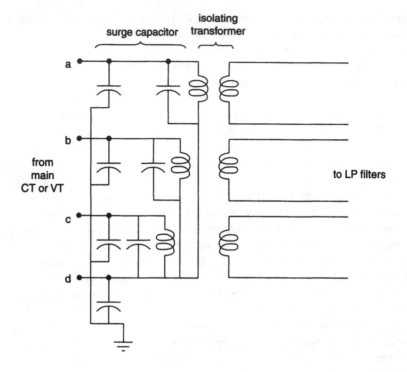

Figure 3.2   Simple surge protective circuit

analogue signal components above a certain frequency (which in turn is related to the digital sampling frequency) be attenuated to avoid errors in subsequent digital processing. It is this 'anti-aliasing' function that is importantly fulfilled by the analogue low-pass filters, which must be designed with a cutoff frequency (f_c) that performs satisfactory signal component rejection above a given frequency. Figure 3.3(*a*) shows the characteristics of an ideal low-pass filter, which transfers signal components of frequencies below the cutoff frequency with zero attenuation, while components above the cutoff frequency are attenuated to zero. The effect of introducing a practical low-pass filter is shown in Figure 3.3(*b*), from which it can be seen that in practice it is not possible to achieve such a pronounced transition between the pass and stop bands.

The dynamic characteristic of the low-pass filters, as well as their steady-state characteristics, are important. Among the more important factors are

(i)   the rise time, which gives an indication of how long it takes the output of a low-pass filter to traverse its final value following a step input

(ii) the overshoot, which indicates by how much the filter output will exceed its steady-state value on the initial response to a unit step input

(iii) the settling time, which is an indication of how long it takes a given filter to settle at its steady-state output value.

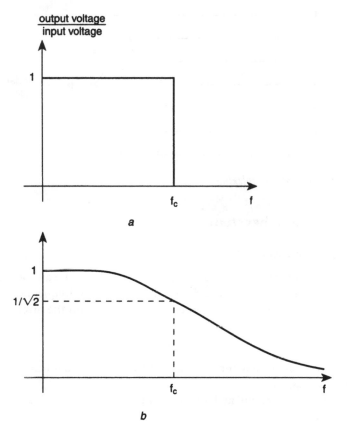

Figure 3.3   Characteristic of low-pass filters; (a) ideal filter response, (b) practical filter response

All the above features play a part in the overall dynamic response of digital relay systems. In particular, in systems where a very high speed decision is required, it is particularly important to ensure that the low-pass filter is designed to have a cutoff frequency that gives an overall performance that is not degraded by long filter delays [6].

3.3.4 Analogue multiplexers [7]

In digital relaying applications, it is usually necessary to use an analogue multiplexer. The concept of multiplexing has its origins in communications engineering. An analogue multiplexer is a device that selects a signal from one of a number of input channels and transfers it to its output channel, thereby permitting the transmission of several signals in a serial manner over a single communication channel. The principles of multiplexing are thus as shown in Figure 3.4, in which the solid-state multiplexer is likened to a multi-terminal rotary switch.

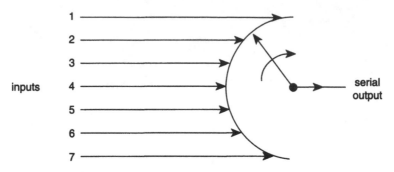

Figure 3.4   Principle of multiplexing

3.4 Conversion subsystem

3.4.1 The sampling theorem [8]

The sampling theorem states that a band-limited signal can be uniquely specified by its sampled values if and only if the sampling frequency is at least twice the maximum frequency component contained within the original signal, i.e.

$$f_s \geqslant 2f_m \tag{3.1}$$

where f_s is the sampling frequency and f_m is the maximum significant frequency within the signal sampled.

The frequency component at half the sampling frequency is known as the Nyquist frequency f_N, or

$$f_N = f_s/2 \tag{3.2}$$

The sampling process is effectively achieved by connecting an analogue signal $f(t)$ to the data acquisition system by means of a fast acting switch, which closes for a very short time but remains open for the rest of the period (Figure 3.5(a)). This operation can be modelled by a multiplier (Figure 3.5(b)), where $f(t)$ is the band-limited analogue signal to be sampled, and $s(t)$ is known as a sampling function. The sampling function is, therefore, made of a train of pulses alternating between a value of $+1$ and 0. It is thus defined as follows:

$$s(t) = \sum_{n=-\infty}^{\infty} \delta(t - nT_s) \tag{3.3}$$

The output of the multiplier $f_s(t)$ is then

$$f_s(t) = f(t) \sum_{n=-\infty}^{\infty} \delta(t - nT_s)$$

$$= \sum_{n=-\infty}^{\infty} f(nT_s)\delta(t - nT_s) \tag{3.4}$$

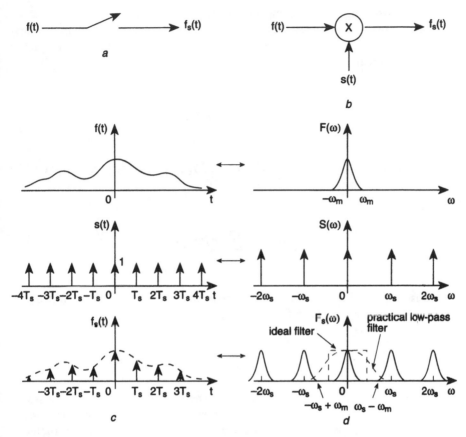

Figure 3.5 Sampling processes; (a) the sampler, (b) representation of the sampler, (c) functions $f(t)$, $s(t)$ and $f_s(t)$, (d) Fourier transforms of $f(t)$, $s(t)$ and $f_s(t)$

Figure 3.5(c) shows the three time functions $f(t)$, $s(t)$ and $f_s(t)$. It can be seen that the sampling function $f_s(t)$ consists of a train of pulses spaced equally by a period equal to the sampling interval T_s. The amplitude of each sample is the same as that of the original signal at the respective sample time nT_s. Therefore, the resulting samples are included within the original signal envelope, as shown in Figure 3.5(c).

To show how and under what conditions $f(t)$ can be uniquely reconstructed from $f_s(t)$, let us examine the spectra of the three time functions $f(t)$, $s(t)$, and $f_s(t)$. This is shown in Figure 3.5(d), in which it is assumed that the function $F(\omega)$ contains no frequency above ω_m. The spectrum of the sampling function is thus given by

$$F_s(\omega) = \omega_s \sum_{n=-\infty}^{\infty} S(\omega - n\omega_s) \tag{3.5}$$

where ω_s is the angular sampling frequency, given by $\omega_s = 2\pi f_s = 2\pi / T_s$.

It will be noted that, in accordance with the convolution theorem, multiplication of two functions in the time domain is equivalent to their convolution in the frequency domain. Therefore,

$$F_s(\omega) = \frac{1}{2\pi} F(\omega) * S(\omega) \qquad (3.6)$$

The convolution of $F(\omega)$ with each pulse of $S(\omega)$ produces $F_s(\omega)$ displaced in frequency steps equal to $n\omega_s$. Thus

$$F_s(\omega) = \frac{1}{T_s} \sum_{n=-\infty}^{\infty} F(\omega - n\omega_s) \qquad (3.7)$$

As illustrated in Figure 3.5(d), it is important to note that there will be no overlap of adjacent parts of $F_s(\omega)$ as long as

$$\omega_s \geqslant 2\omega_m \qquad (3.8)$$

If this condition is fulfilled, the passing of the signal $f_s(t)$ through a low-pass filter (see Figure 3.5(d)) with a bandwidth such that its cutoff frequency ω_c is given by eqn. 3.9, results in a perfect digital reconstruction of the analogue signal $f(t)$.

$$\omega_m \leqslant \omega_c \leqslant \omega_s - \omega_m \qquad (3.9)$$

3.4.2 Signal aliasing error [9]

If the sampling rate is chosen so that the sampling frequency is less than twice the maximum significant frequency contained in the original signal, i.e. eqn. 3.1 is not satisfied, then it can be seen from Figure 3.5(d) that there will be an overlap between adjacent parts of $F_s(\omega)$. This leads to what is commonly known as an 'aliasing error', which in turn causes an error in the analysis as a result of the difficulty in distinguishing between low- and high-frequency components. In other words, if the sampling rate does not satisfy eqn. 3.1, a low-frequency component that does not actually exist in the original signal, would nevertheless be apparent within the sampled signal. Figure 3.6 gives an illustration of the aliasing phenomenon, in which the dotted line represents a low frequency that does not actually exist in the original signal.

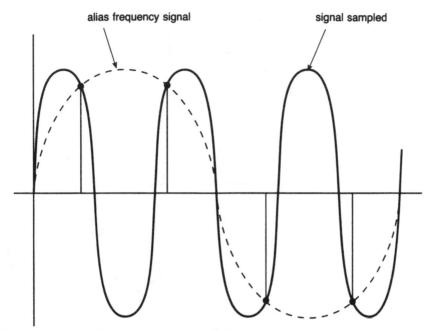

alias frequency signal   signal sampled

Figure 3.6 Simple illustration of aliasing phenomenon

3.4.3 Sample and hold circuit [10–12]

If the sampling switch of Figure 3.5(a) is replaced by a switch-capacitor combination, each sample can be stored or held until the next sample is taken. This process allows sufficient time to elapse for the subsequent process of analogue to-digital conversion to be completed. Figure 3.7 shows the principles of a simple sample-and-hold circuit applied to the signal derived from the output of the signal-conditioning subsystem (see Figure 3.1).

Figure 3.7(a) shows a basic sample-and-hold circuit. The input is the analogue signal $f(t)$, which is sampled at a rate of $1/T_s$. Sampling is controlled by the control waveform V_c, which closes and opens the switch. During the closing time T_c, the capacitor is charged to the value $f(t)$, while during the hold time $T_H = T_s - T_c$, the capacitor holds the sample value. The process of sample and hold is best understood by referring to Figures 3.7(b), (c) and (d). There are a variety of sample-and-hold circuits available in i.ttegrated-circuit (IC) form. However, virtually all commercially available devices work on the principle outlined, their performance being defined largely in terms of accuracy and sampling rates.

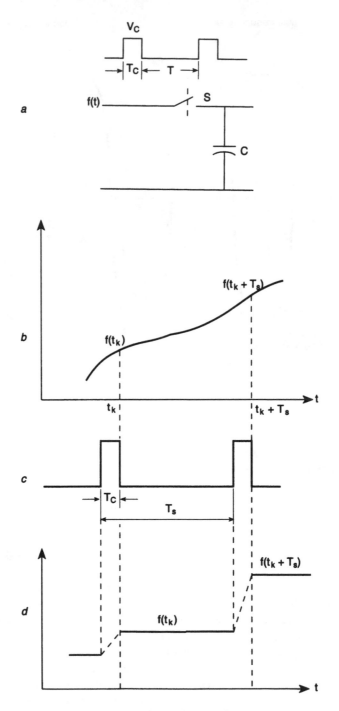

Figure 3.7   Sample and hold circuit principle (a) switching circuit, (b) analogue input, (c) control waveform, (d) circuit output

3.4.4 *Digital multiplexing* [13–15]

A digital multiplexer is a network with a number of input ports channelled to a single output. The input to these ports consists of digital words of information of one or more bits.

Figure 3.8 shows a general representation of a digital multiplexer that allows one of its three inputs to pass to the output side, such that

$$\bar{Y} = 0 \qquad \text{if } [A, B] = [0, 0]$$
$$\bar{Y} = I1 \qquad \text{if } [A, B] = [0, 1]$$
$$\bar{Y} = I2 \qquad \text{if } [A, B] = [1, 0] \qquad\qquad (3.10)$$
$$\bar{Y} = I3 \qquad \text{if } [A, B] = [1, 1]$$

where

$$\bar{Y} = [Y_0, Y_1, \ldots Y_n] \text{ is the output}$$
$$\overline{I1} = [Y_{01}, Y_{11}, \ldots Y_{n1}] \text{ is the first input}$$
$$\overline{I2} = [Y_{02}, Y_{12}, \ldots Y_{n2}] \text{ is the second input}$$
$$\overline{I3} = [Y_{03}, Y_{13}, \ldots Y_{n3}] \text{ is the third input}$$

Figure 3.9 shows a typical logic diagram for a two-input-to-one-output multiplexer. The input word $\overline{Y1}$, which consists in this case of four bits (Y_{01}, Y_{11}, Y_{21}, Y_{31}) is transferred to the output $\bar{Y}$ if the control signal (address) A is low. On the other hand, when the address signal A is high, then the input $\overline{Y2}$ is transferred to the output $\bar{Y}$.

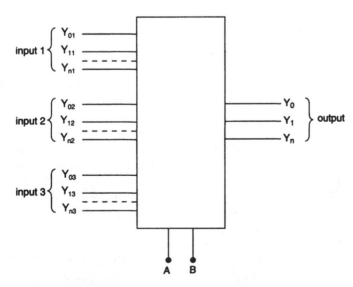

Figure 3.8   Digital multiplexer arrangement

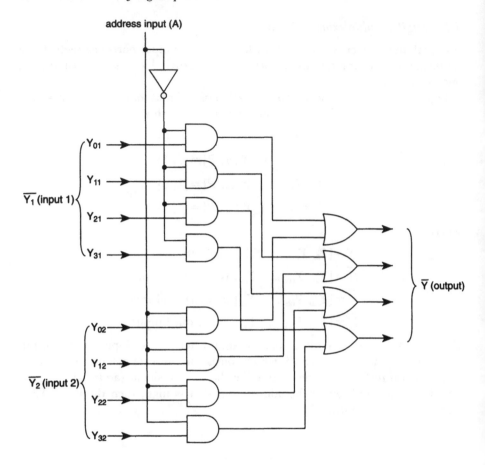

address input (A)

Y_{01}

Y_{11}

$\overline{Y_1}$ (input 1)

Y_{21}

Y_{31}

Y_{02}

Y_{12}

$\overline{Y_2}$ (input 2)

Y_{22}

Y_{32}

$\overline{Y}$ (output)

Figure 3.9   Logic circuit for two input/one output multiplexer

3.4.5 *Digital-to-analogue conversion* [12–16]

Consider first the method of converting a digital representation to analogue form. The reason for this is that D/A converters are frequently used in analogue-to-digital converters. The principles involved are well illustrated by considering the circuit of Figure 3.10. This may be used to convert a four-bit binary parallel digital word $W_3W_2W_1W_0$, where any W_i is either 0 or 1, to an analogue voltage that is proportional to the binary number represented by the digital word. The logic voltages that represent any individual bit W_i are, as a matter of fact, not connected to the converter but rather are used to control the switches S_0, S_1, S_2 and S_3. Therefore if $W_i = 1$, then S_i is connected to V_R while if $W_i = 0$, then S_i is connected to ground.

The values of the circuit elements are chosen so that successive resistors on the input side are related in powers of two, and the individual resistors are inversely proportional to the numerical significance of the appropriate binary digit (Figure 3.10).

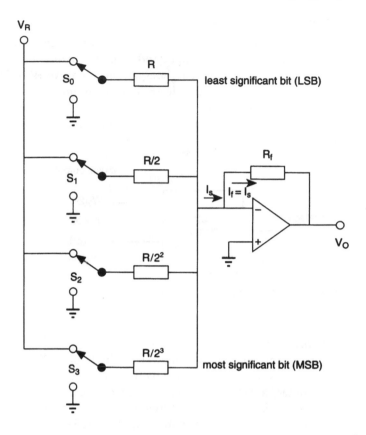

Figure 3.10   Basis of simple digital-to-analogue converter arrangement

Remembering that the input impedance of the operational amplifier is very high, the currents I_s and I_f are roughly equal to each other. The relationship between the output voltage of the operational amplifier (V_o) and the digital input can be found as follows:

$$I_S = (2^3 W_3 + 2^2 W_2 + 2^1 W_1 + 2^0 W_0) V_R / R \qquad (3.11)$$

$$V_o = -I_f R_f = -I_s R_f$$

$$= -(2^3 W_3 + 2^2 W_2 + 2^1 W_1 + 2^0 W_0) V_R R_f / R \qquad (3.12)$$

It is clear from eqn. 3.12 that the output signal voltage V_o is directly proportional to the binary number $W_3 W_2 W_1 W_0$. For illustrative purposes assume that the binary input array corresponds to 15 (decimal), i.e. $W_3 W_2 W_1 W_0 = 1111$; in this case the output is proportional to $(2^3 + 2^2 + 2^1 + 2^0) = 15$ as required. Similarly, an input word of 0011 would give an output proportional to $(0 + 0 + 2^1 + 2^0)$, corresponding to the input number 3.

3.4.6 Analogue-to-digital conversion

The block schematic of an analogue-to-digital converter is shown in Figure 3.11, where V_a is the analogue voltage input and $\bar{W}$ is the digital output array. For simplicity, a four-bit word converter giving an output array $W_3W_2W_1W_0$ is shown. In the field of digital integrated electronics, many methods have been developed for converting analogue signals to digital form, the most commonly used being counter-controlled converters, dual-slope converters and parallel-comparator converters.

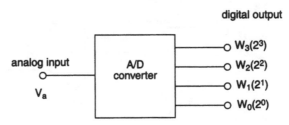

Figure 3.11   Block diagram of analogue-to-digital converter

3.4.6.1 Counter-controlled converter

This technique represents one of the simplest methods of A/D conversion. Figure 3.12 shows the basic arrangement involved. It consists of three main components: a counter, a D/A converter and an analogue comparator. Its operating principles may be summarised as follows. The counter is set to zero at the beginning of each cycle, which causes the output of the D/A converter V_c to be zero. This latter voltage is compared with the input voltage V_i fed by the sample-and-hold circuit, using the comparator whose output will be either 1 or 0 depending on the relative magnitudes of V_i and V_c. If $V_i > V_c$ then the comparator output will be 1, which is used to enable the AND gate. This in turn allows the clock pulses to enter the counter.

Each pulse entered in the counter causes the output voltage of the A/D converter to increase by a single step of 1 V, as shown in Figure 3.12(*b*). As soon as V_c becomes greater then V_i, i.e. $V_i < V_c$, the output of the comparator will be zero. This disables the AND gate, and the clock pulses are thereafter prevented from reaching the counter, which then stops. The output of the A/D converter is then read from the output terminals of the counter. The example illustrated in Figure 3.12 involves an analogue input voltage that is marginally above 5 quantum levels, and is identified as the equivalent binary number 0101. The resolution of the counter is controlled by the gain on the D/A converter, while the dynamic range is largely dictated by the number of binary stages within the counter. In relay applications quantum levels between 5 μV and 300 μV are fairly common, the precise level being a function of dynamic range requirements. This type of counter is useful in some applications, but has limited speed for a given resolution.

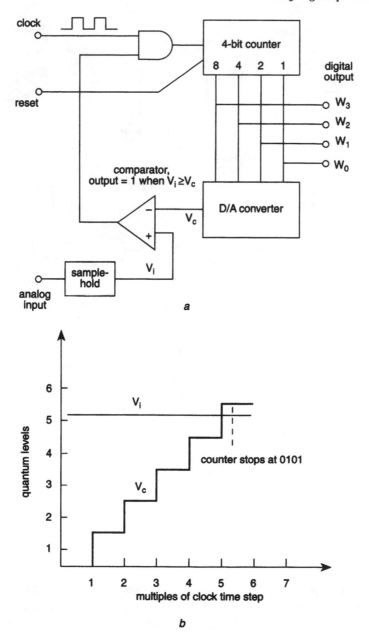

Figure 3.12   Counter-controlled converter (a) circuit arrangement (b) illustrative response

3.4.6.2 Dual-slope converter

This converter basically consists of an analogue integrator, a comparator and a counter. Its operating principle is best understood by reference to Figure 3.13.

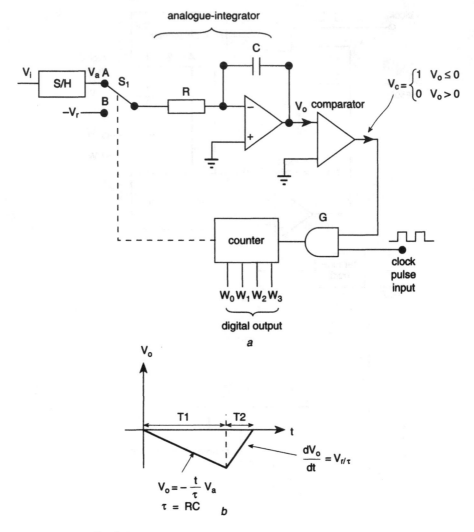

Figure 3.13   Dual-slope converter

At time $t=0$, the switch S_1 is connected to A, and the sample hold voltage V_a is applied to the analogue integrator, which integrates V_a for a fixed time T_1. After that the switch S_1 is thrown to B, which in turn connects a negative reference voltage $-V_r$ to the integrator. Therefore the reference voltage will be integrated and the output of the integrator starts to move in the positive direction (Figure 3.13(b)). As soon as the integrator output voltage V_c reaches zero, the gate G is disabled, and this in turn stops the clock pulses from reaching the counter, thereby stopping the count. The number of clock pulses admitted to the counter is thus proportional to the magnitude of the input voltage V_a which is thereby converted to a digital representation. In practice, the reset voltage V_r is arranged to stop the count quickly. This type of converter is characterised by a high accuracy, but it does have a relatively slow conversion rate, which limits its application in some digital relaying devices.

3.4.6.3 Parallel-comparator converter

Figure 3.14 illustrates a three bit parallel-comparator A/D converter. It consists of comparators C_1 to C_7, a register of seven flipflops and a decoder.

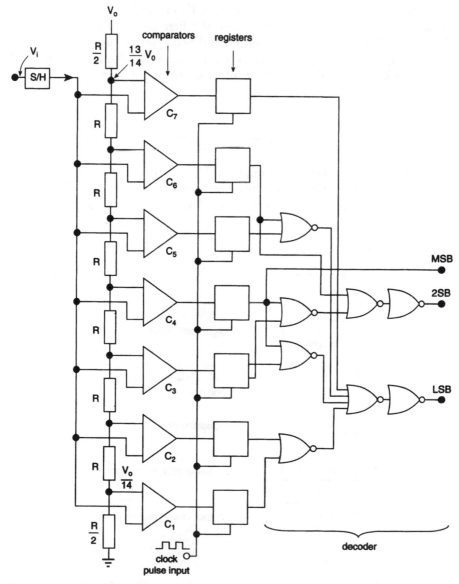

Figure 3.14   Parallel-comparator converter

The input ranges from 0 to V_o and is divided into eight reference segments, six of which have a value of $V_o/7$ the two end segments having a value of $V_o/14$ each. If an input voltage V_i, with magnitude ranging from 0 to $V_o/14$, is applied to the converter, then the outputs of all comparators C_1–C_7 are set to logical 0. The digital representation of this at the output of the converter is seen to be 000. This is equivalent, as required, to a zero analogue voltage. It will be noted that an error of $V_o/14$ is introduced on account of the input being equal to one level of resolution ($V_o/14$). This is often called a 'quantisation' error.

If the input voltage magnitude is between $V_o/14$ and $3V_o/14$, then the output of all the comparators is zero, except the first one which will be 1, i.e. $C_1C_2C_3C_4C_5C_6C_7 = 1000000$. The output of the comparators is transferred to the register flipflops at the occurrence of a clock pulse and the register output is converted by the decoder to a three-bit unipolar binary code.

3.5 Digital relay subsystem [5, 6, 17]

The digital relay subsystem comprises both hardware and software. The hardware largely consists of a central processor unit (CPU), memory, data input and output (I/O). The software is influenced by two major factors. The

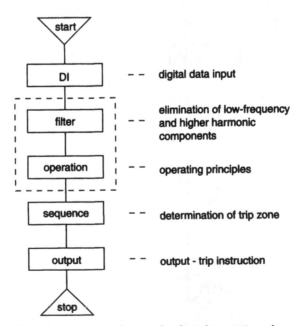

Figure 3.15   Flow chart for the software of a digital protective relay

first of these is the operating principles and performance required, which either leads to the development of a special algorithm or the implementation of an existing one. This factor greatly affects the determination of the sampling frequency, type of hardware structure and the data input hardware system.

The second factor is the digital filtering. Subharmonics, as well as high harmonic components, can cause false tripping, failure to trip and variation in protective relay performance. The operating principle and digital filtering must in general provide for a wide application range and requirements relating to speed of response to system faults, the latter being influenced by the necessary computation time.

Figure 3.15 shows an example of a flow-chart for the software of a typical digital protective relay. The algorithms used and the software required vary significantly according to the application, and much of the work of later chapters will be concerned with specific algorithms for meeting a variety of protection performance and application requirements.

3.6 The digital relay as a unit

There is an inevitable tendency for relay manufacturers to develop standardised hardware, which can be used in conjunction with suitably developed software to meet a variety of production requirements and applications. Figure 3.16 shows a block diagram of a standardised digital relay unit, and Table 3.1 gives a typical basic specification of such a unit.

Table 3.1   Specification of a typical standardised digital relay unit [5]

Data acquisition		Sampling frequency Data length	720 600 sample/s 12 bit
Data processing	CPU	Device word length language	bipolar 16 bit assembler
Memory		IC memory	
Tap setting		Nonvolatile IC memory (up to 256 words)	

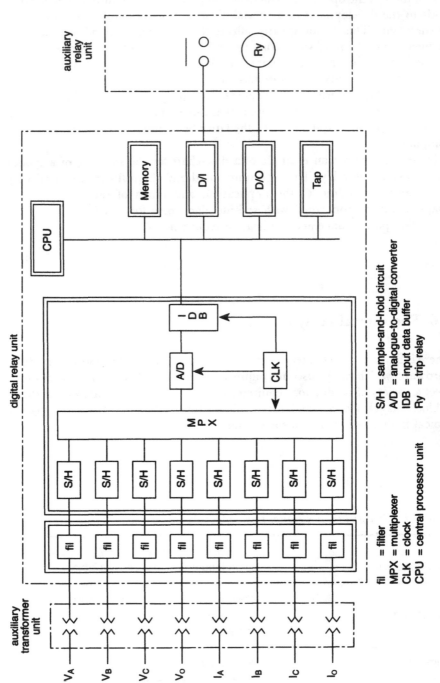

fil   = filter
MPX = multiplexer
CLK = clock
CPU = central processor unit

S/H = sample-and-hold circuit
A/D = analogue-to-digital converter
IDB = input data buffer
Ry   = trip relay

Figure 3.16   Digital relay unit

3.7 References

1 WRIGHT, A.: 'Current transformers—their transient and steady state performance' (Chapman & Hall, 1968)
2 STALWESKI, A., and WELLER, G. E.; 'Novel capacitor divider voltage sensors for high-voltage transmission systems', *Proc. IEE*, 1979, **126**, pp. 1186–1195
3 'Computer relaying', IEEE Tutorial course 79 EH01487-PWR, 1979
4 GILCHRIST, G.B., ROCKEFELLER, G.D., and UDREN, E.A.: 'High speed distance relaying using a digital computer, part I—system description', *IEEE Trans.* 1972, **PAS-91**, pp. 1235–1243
5 SEKINE, Y., HATATA, M., and YOSHIDA, T.: 'Recent advances in digital protection', *Elect. Power & Energy Syst.*, 1984, **6**, pp. 181–191.
6 JOHNS, A.T., and WALKER, E.P.: 'Genesis and evolution of advanced protection', *Power Eng. J.* 1991, **5** (4), pp. 177–187
7 CAHILL, J.J.: 'Digital and microprocessor engineering' (Ellis Horwood, 1982)
8 MAYHAN, R.J.: 'Discrete-time and continuous-time linear systems' (Addison Wesley, 1984)
9 NATARAJAN, N.A.T.: 'Discrete-time signals and systems' (Reston, 1983)
10 TAUB, H., and SCHILLING, D.: 'Digital integrated electronics' (McGraw-Hill, 1977)
11 STANIER, B.J.: 'Electronic and integrated circuits' (Adam Hilger, 1985)
12 SCHILLING, D., and BELVE, B.: 'Electronic circuits—discrete and integrated' (McGraw-Hill, 1979)
13 *ibid*, 3rd edition (McGraw-Hill, 1989)
14 BOOTH, T.L.: 'Digital network and computer systems' (2nd edition), (John Wiley, 1978).
15 HILL, F.J.: 'Introduction to switching theory and logical design' (John Wiley, 1981)
16 MALVINO, A.P.: 'Digital computer electronics' (McGraw-Hill, 1977)
17 MAKINO, J., and MIKI, Y.: 'Study of operating principle and digital filters for protective relays with digital computers', IEEE PES Winter Power Meeting, New York, January 1975, 75CHO990-2 PWR, Paper No. C75 197-9, pp. 1–8.

Chapter 4

Sinusoidal-wave-based algorithms

4.1 Introduction

The algorithms covered in this Chapter assume that the post-fault current and voltage waveforms are sinusoidal. This assumption is not, of course, generally valid, particularly when EHV or UHV network applications are involved. However, in practice, the signals processed are often prefiltered and, in lower voltage distribution systems in particular, the waveforms often very quickly attain a nominally sinusoidal form. Historically, algorithms developed for use in applications where the signals processed are nominally sinusoidal were the first to emerge [1, 2]. Most of the early work involved applying the technique to the calculation of transmission-line fault-loop impedances. However, the methods are equally applicable to determining the magnitude and phase of relaying currents for differential protection of lines and plant.

All sinusoidal-based algorithms are designed to predict either the peak or squared peak of the compared waveforms. They may be loosely classified into two broad groups: those that use sample and first derivative (or first and second derivative), and those that use two or three samples to predict the peak or squared peak values.

4.2 Sample and first-derivative method [1, 2]

4.2.1 Basic formulation

When a waveform is assumed to be perfectly sinusoidal, its peak can be predicted from any of its samples. Consider for example two signal measurands, $s_1(t)$ and $s_2(t)$ say, which take the form of eqns. 4.1 and 4.2 respectively.

$$s_1(t) = V_1 \sin \omega_0 t \tag{4.1}$$

$$s_2(t) = V_2 \sin(\omega_0 t + \theta) \tag{4.2}$$

where V_1 and V_2 are the peak values of the signal waveforms. By taking the time derivative of eqn. 4.1, we obtain

$$s_1'(t) = \omega_0 V_1 \cos \omega_0 t \tag{4.3}$$

(where, for brevity, $s_1'(t)$ is written for $\dfrac{d}{dt} [s_1(t)]$).

When eqns. 4.1 and 4.3 are combined together, we obtain eqn. 4.4, which defines the peak values of relaying signal s_1t.

$$V_1^2 = s_1(t)^2 + \left(\frac{s_1'(t)}{\omega_0}\right)^2 \tag{4.4}$$

Similar equations to eqns. 4.1–4.3 can be derived to describe the peak value of the second relaying waveform $s_2(t)$, i.e.

$$V_2^2 = s_2(t)^2 + \left(\frac{s_2'(t)}{\omega_0}\right)^2 \tag{4.5}$$

Determination of measured impedance
In distance protection, it is necessary to measure the apparent impedance of the fault loop using signal measurands proportional to the voltage and current monitored via the transformers located at the line ends. In this case, the relaying signals can be conveniently described by eqns. 4.6 and 4.7:

$$s_1(t) = v = V \sin \omega_0 t \tag{4.6}$$

$$s_2(t) = i = I \sin(\omega_0 t + \theta) \tag{4.7}$$

In what follows, we will assume for brevity and convenience that the methods described are applied to distance protection, although, as mentioned previously, they may equally be applied to differential protection. The basic measurands will therefore be assumed to take the form of eqns. 4.6 and 4.7.

The measured impedance can be calculated from the magnitude and relative phase of the voltage and current measurands given in eqns. 4.6 and 4.7. The magnitude of the impedance is simply calculated as $|Z| = V/I$ which, with the basic measurands defined in eqns. 4.6 and 4.7 is given by

$$|Z| = \{[v^2 + (v'/\omega_0)^2]/[i^2 + (i'/\omega_0)^2]\}^{1/2} \tag{4.8}$$

The argument (or phase) of the measured impedance (θ) is equal to the angular difference between the voltage and current signal. Thus, the argument of the measured impedance is given by eqn. 4.9:

$$\theta = \theta_V - \theta_I \tag{4.9}$$

where from eqns. 4.6 and 4.7, $\theta_V = \omega_0 t$ and $\theta_I = \omega_0 t + \theta$.

The values of these angles can also be determined from a waveform sample and its derivative. This is done by re-writing eqns. 4.6 and 4.7 in terms of θ_V and θ_I, which in turn results in

$$v = V \sin \theta_V \tag{4.10}$$

$$v' = \omega_0 V \cos \theta_V \tag{4.11}$$

Eqns. 4.10 and 4.11 thus give

$$\theta_V = \tan^{-1}(v\omega_0/v') \tag{4.12}$$

Similarly, θ_I can be found as

$$\theta_I = \tan^{-1}(i\omega_0/i') \tag{4.13}$$

Substituting eqns. 4.12 and 4.13 into 4.9, we thus obtain the argument of the measured impedance in terms of the instantaneous values of the measurands v and i, together with their derivatives:

$$\theta = \tan^{-1}(v\omega_0/v') - \tan^{-1}(i\omega_0/i') \tag{4.14}$$

4.2.2 Calculation of an approximation to the signal derivatives

We have shown in the previous Section that the magnitude ($|Z|$) and angle (θ) of the impedance depends on sample values as well as the derivative of the current and voltage measurands. Therefore, it is essential to know how to determine the required derivatives from measured samples. This is achieved by using the numerical techniques given in Chapter 2, which involves using a series of expressions for forward, central or backward differences. However, for real-time implementation, it is often advantageous to use backward differences because existing samples can be used for immediate derivative calculation.

By substituting a particular voltage sample v_k and the corresponding voltage derivative v'_k for f_k and f'_k, respectively, in eqn. 2.27, we obtain

$$v'_k \approx \frac{1}{h}\left(\nabla + \frac{1}{2}\nabla^2 + \ldots\right)v_k \tag{4.15}$$

where $h = \Delta t$ is sampling time interval and

$$\nabla v_k = v_k - v_{k-1}$$
$$\nabla^2 v_k = \nabla v_k - \nabla v_{k-1}$$
$$= v_k - 2v_{k-1} + v_{k-2}$$

Using only the first term in eqn. 4.15 leads to

$$v'_k \approx \frac{1}{h}(v_k - v_{k-1}) \tag{4.16}$$

Using the first two terms likewise gives

$$v'_k \approx \frac{1}{2h}(3v_k - 4v_{k-1} + v_{k-2}) \tag{4.17}$$

Similar equations can be used to estimate a current derivative (i'). It will be noted that, according to eqn. 4.16, it is possible to calculate the derivative immediately after two samples become available, whereas the use of eqn. 4.17 requires that a third sample be available. The 'three sample' version does of course give a generally more accurate estimate of the differential but requires more computation.

4.2.3 Error analysis

To determine the error involved in numerically determining derivatives, let us assume that the sinusoidal waveforms are sampled at discrete time intervals $h = \Delta t$, with the actual sampling times being t_k, t_{k+1}, ... at which the

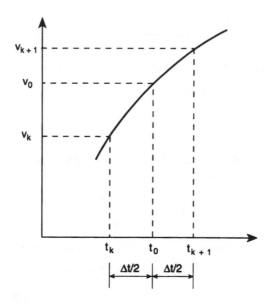

Figure 4.1   Calculation of voltage v_0 halfway between samples taken at times t_k and t_{k+1}

corresponding, voltage samples are v_k, v_{k+1}, ... Let t_0 be a time halfway
between t_k and t_{k+1}, as shown in Figure 4.1. Then it follows that

$$v_0 \approx \frac{1}{2}\,(v_k + v_{k+1})$$

and   (4.18)

$$v_0' \approx \frac{\Delta v}{\Delta t} = \frac{1}{\Delta t}\,(v_{k+1} - v_k)$$

Given that we are dealing with a sinusoidal waveform, the sampled value at
time t_k can be written as $v_k = V \sin \omega_0 t_k$, which in turn is given by

$$v_k = V \sin \omega_0(t_0 - \Delta t/2) \qquad\qquad (4.19)$$

The voltage sampled one time interval later (v_{k+1}) is likewise given by

$$v_{k+1} = V \sin \omega_0(t_0 + \Delta t/2) \qquad\qquad (4.20)$$

By substituting eqns. 4.19 and 4.20 into eqns. 4.18 we obtain

$$v_0 \approx V \sin(\omega_0 t_0)\,\cos(\omega_0 \Delta t/2) \qquad\qquad (4.21)$$

$$v_0' \approx [2V \cos(\omega_0 t_0)\,\sin(\omega_0 \Delta t/2)]/\Delta t \qquad\qquad (4.22)$$

Because the value of $\omega_0 \Delta t/2$ is small, the sine and cosine terms can be readily
approximated by

$$\cos \frac{\omega_0 \Delta t}{2} \approx 1 - \frac{(\omega_0 \Delta t/2)^2}{2} \qquad\qquad (4.23)$$

$$\sin \frac{\omega_0 \Delta t}{2} \approx \frac{\omega_0 \Delta t}{2} \tag{4.24}$$

By substituting eqns. 4.23 and 4.24 into 4.21 and 4.22, we obtain

$$v_0 \approx V \sin(\omega_0 t_0) [1 - (\omega_0 \Delta t)^2/8] \tag{4.25}$$

and

$$v_0' \approx \omega_0 V \cos \omega_0 t_0 \tag{4.26}$$

The estimated value of the peak of the voltage signal using sampled data at t_0 (V_0) can be found using eqn. 4.4:

$$V_0^2 = v_0^2 + \left(\frac{v_0'}{\omega_0}\right)^2 \tag{4.27}$$

When this equation is combined with eqns. 4.25 and 4.26, we obtain

$$V_0^2 \approx V^2[1 - (\omega_0^2 t^2 \sin^2 \omega_0 t_0)/4]$$

or

$$V_0 \approx V[1 - (\omega_0^2 \Delta t^2 \sin^2 \omega_0 t_0)/8] \tag{4.28}$$

It follows that the percentage error ($\varepsilon\%$) associated with the estimated value of the peak of the voltage waveform will be

$$\varepsilon\% = \frac{V_0 - V}{V} \times 100$$

$$\varepsilon\% = -\frac{\omega_0^2(\Delta t)^2}{8} \sin^2 \omega_0 t_0 \times 100 \tag{4.29}$$

Eqn. 4.29 indicates that the error in the estimated peak value is a function of the system frequency ω_0, the sampling interval Δt and the time at which the estimation takes place. For a sampling interval Δt of 0.5 ms, the error is seen to be a maximum when the estimation takes place at the extreme of the waveform, i.e. at $\omega_0 t_0 = \pm n \frac{\pi}{2}$, $n = 1, 2, \ldots$ On a 50 Hz system, this error is of the order of 0.3% (0.45% for 60 Hz systems).

4.2.4 Practical considerations

4.2.4.1 Decaying DC components

Practically, the current and voltage waveforms that immediately follow power system faults are never pure sinusoids. This is particularly true for current waveforms, which often contain a relatively large and slowly decaying DC component. It follows that, in applying the above derived algorithm, care must be taken in dealing with any DC offset components in the current signals, which in practice can be transferred in full to the digital processing stage. This is particularly so where the input circuit comprises an iron-cored transformer with resistive burden.

The DC offset in the current signal can be reduced by using a mimic impedance in the current transformer secondary, having an X/R ratio close to that of the primary circuit. However, in practice it is not possible to match the X/R ratio of the mimic to that of the power circuit for the complete range of possible fault conditions, and some error is therefore caused by exponentially decaying current and voltage components. Nevertheless, this technique can greatly increase the range of applicability of techniques that assume purely sinusoidal measurands.

4.2.4.2 Noise and digital smoothing

Another practical point of interest is that related to the error introduced by noise in the input signals. Such errors can be reduced using digital smoothing of sampled data. To avoid large delays caused by the process of digital smoothing, three samples may be used with a first degree (straight line) approximation. To meet these requirements, eqns. 2.56–2.58 can be used to calculate the values of digitally smoothed samples. When this is done, the smoothed values of sampled values corresponding to sampling instants $k-1$, k and $k+1$ are substituted into the basic algorithm described. Denoting the smoothed values of (say) a voltage by upper case letters, the values substituted would take the form of eqns. 4.30.

$$V_{k-1} = (5v_{k-1} + 2v_k - v_{k+1})/6$$
$$V_k = (v_{k-1} + v_k + v_{k+1})/3 \qquad (4.30)$$
$$V_{k+1} = (-v_{k-1} + 2v_k + 5v_{k+1})/6$$

Eqns. 4.30 use forward, central and backward points, respectively. However, in most applications of the algorithm it is preferable to use the backward expression, because a smoothed value can be generated from the data already in hand, thus speeding up the overall response.

4.3 First- and second-derivative method [3, 4]

Algorithms of this type generally reduce errors arising from subnormal frequencies, as well as those due to slowly decaying DC transients. In essence, they represent a refinement of the above detailed basic algorithms and play an important part in applications where significant periodic and aperiodic components are present in the signal waveforms.

4.3.1 Mathematical formulation

As previously, let the voltage and current waveforms be denoted by eqns. 4.6 and 4.7, i.e. $v = V \sin \omega_0 t$ and $i = I \sin(\omega_0 t + \theta)$.

Taking the first and second derivative with respect to time we obtain for the voltage signal

$$v' = \omega_0 V \cos \omega_0 t \qquad (4.31)$$

$$v'' = -\omega_0^2 V \sin \omega_0 t \qquad (4.32)$$

Combining these two equations results in an equation for the square of the peak of the assumed sinusoidal voltage:

$$V^2 = \frac{1}{\omega_0}\left[(v')^2 + \left(\frac{v''}{\omega_0}\right)^2\right] \tag{4.33}$$

The corresponding equation for determining an approximation to the peak of the current is likewise

$$I^2 = \frac{1}{\omega_0}\left[(i')^2 + \left(\frac{i''}{\omega_0}\right)^2\right] \tag{4.34}$$

The square of the magnitude of the measured impedance is

$$|Z|^2 = \frac{V^2}{I^2} = \frac{(v')^2 + \left(\dfrac{v''}{\omega_0}\right)^2}{(i')^2 + \left(\dfrac{i''}{\omega_0}\right)^2} \tag{4.35}$$

The argument of the measured impedance can be determined by using procedures that are similar to those given in the previous section. Therefore, from eqns. 4.31 and 4.32, the voltage angle is

$$\theta_V = \omega_0 t = \tan^{-1}\frac{v''/\omega_0^2}{-v'/\omega_0}$$

$$= -\tan^{-1}\left(\frac{v''}{\omega_0 v'}\right)$$

and that of the current is

$$\theta_I = -\tan^{-1}\left(\frac{i''}{\omega_0 i'}\right)$$

It follows that the argument of the measured impedance angle (θ) is given by

$$\theta = \tan^{-1}\left(\frac{i''}{\omega_0 i'}\right) - \tan^{-1}\left(\frac{v''}{\omega_0 v'}\right)$$

First and second derivatives are commonly determined for use in this algorithm by using divided differences (see Chapter 2). This is done by substituting the voltage v and Δt for the variables f and h, respectively, into eqns. 2.33 and 2.34, in which case the following relationships are used:

$$v'_k = \frac{1}{2\Delta t}(v_{k+1} - v_{k-1}) \tag{4.36}$$

$$v''_k = \frac{1}{(\Delta t)^2}(v_{k+1} - 2v_k + v_{k-1}) \tag{4.37}$$

where Δt is the sampling interval, and $k+1$, k and $k-1$ are subscripts referring to a set of consecutive samples.

From eqns. 4.36 and 4.37, it can be seen that the main advantage of this method is that a constant DC component has no effect since it is cancelled out from all difference expressions. The effect of the non-zero frequency components due to any exponentially decaying offsets is consequently minimised, and no mimic burden or related technique is required in many applications. However, the disadvantage associated with this algorithm is that higher-frequency components, due in particular to fault-induced travelling waves, can produce significant errors in the first- and second-difference equations, thereby causing large errors in successive impedance estimates. A solution to this problem, which is successful in many applications, is to use the smoothed form of sample and first derivative version of the algorithm for processing the voltage signal and the first and second derivative technique for the current. In this way, maximum overall "noise" rejection can be obtained, thereby improving the reliability of successive impedance estimates.

4.4 Two-sample technique [5]

In previous sections, the calculation of the magnitude and phase of the voltage and current waveforms was achieved using derivatives. An alternative, which avoids the need for finding differentials, involves manipulating two samples taken at discrete instants of time from the signal waveforms.

4.4.1 Prediction of values of peak (or magnitude) of signal waveforms

Let v_k, v_{k+1} be voltage samples measured at times t_k, t_{k+1} respectively, and let Δt be the sampling time interval. Then

$$v_k = V \sin \omega_0 t_k \tag{4.38}$$

$$v_{k+1} = V \sin \omega_0 t_{k+1} = V \sin \omega_0 (t_k + \Delta t)$$

$$v_{k+1} = V \sin \omega_0 t_k \cos \omega_0 \Delta t + V \cos \omega_0 t_k \sin \omega_0 \Delta t \tag{4.39}$$

Substituting eqn. 4.38 into 4.39 and simplifying results in

$$\frac{v_{k+1} - v_k \cos \omega_0 \Delta t}{\sin \omega_0 \Delta t} = V \cos \omega_0 t_k \tag{4.40}$$

Adding the squares of eqns. 4.38 and 4.40, and noting that $\sin^2 x + \cos^2 x = 1$, we obtain an equation for the square of the peak voltage:

$$V^2 = \frac{v_k^2 + v_{k+1}^2 - 2v_k v_{k+1} \cos \omega_0 \Delta t}{(\sin \omega_0 \Delta t)^2} \tag{4.41}$$

The corresponding equation for the peak of the current is simply:

$$I^2 = \frac{i_k^2 + i_{k+1}^2 - 2i_k i_{k+1} \cos \omega_0 \Delta t}{(\sin \omega_0 \Delta t)^2} \tag{4.42}$$

It can be seen that the values $\omega_0\Delta t$ are fixed for any given nominal system angular frequency and sampling interval Δt. They therefore appear as constants in the evaluation of the peak values, which, in effect, simply weight the sampled values used in the algorithm.

4.4.2 Determination of phase angle between waveforms

The phase angle between voltage and current waveforms can be calculated using the above determined values of waveform magnitude (V and I) together with the measured voltage and current samples.

Let θ be the phase angle between the voltage and current waveform such that the current sample i_k is expressed as:

$$i_k = I\,\sin(\omega_0 t_k + \theta)$$

or

$$i_k = I\,\sin\,\omega_0 t_k\,\cos\,\theta + I\,\cos\,\omega_0 t_k\,\sin\,\theta \qquad (4.43)$$

The current sample i_{k+1} (taken at a time $t_k + \Delta t$) is likewise obtained in the form

$$i_{k+1} = I[\sin\,\omega_0 t_k\,\cos\,\omega_0\Delta t + \cos\,\omega_0 t_k\,\sin\,\omega_0\Delta t]\,\cos\,\theta$$
$$+ I[\cos\,\omega_0 t_k\,\cos\,\omega_0\Delta t - \sin\,\omega_0 t_k\,\sin\,\omega_0\Delta t]\,\sin\,\theta \qquad (4.44)$$

From eqns. 4.38 and 4.40 we have

$$\sin\,\omega_0 t_k = \frac{v_k}{V} \qquad (4.45)$$

$$\cos\,\omega_0 t_k = \frac{v_{k+1} - v_k\,\cos\,\omega_0\Delta t}{V\,\sin\,\omega_0\Delta t} \qquad (4.46)$$

Eqns. 4.43 to 4.46 can be manipulated to obtain a number of algorithmic equations, the simplest of which is

$$\theta = \cos^{-1}\left\{ \frac{i_k v_k + i_{k+1} v_{k+1} - (i_k v_{k+1} + i_{k+1} v_k)\,\cos\,\omega_0\Delta t}{IV\,\sin^2\,\omega_0\Delta t} \right\} \qquad (4.47)$$

The above equation shows that the phase angle θ can be determined from a pair of voltage samples and current samples. Despite the apparent complexity of its form, the above equation is in fact relatively easy to evaluate, given that, as part of the overall determination of the measured impedance, the peak values of the voltage and current measurands (V and I) must be evaluated anyway.

4.5 Three-sample technique [6]

This algorithm predicts the peak of the voltage and current measurands by using three successive samples. It also calculates the resistive and reactive parts of the impedance. This contrasts with the previously described algorithms, which produce estimates of the magnitude and argument of the measured impedance.

Let v_k, v_{k+1}, v_{k+2} be voltage samples measured at times t_k, t_{k+1} and t_{k+2}, then

$$v_k = V \sin \omega_0 t_k \qquad (4.48)$$

$$v_{k+1} = V \sin(\omega_0 t_k + \omega_0 \Delta t) \qquad (4.49)$$

$$v_{k+2} = V \sin(\omega_0 t_k + 2\omega_0 \Delta t) \qquad (4.50)$$

If i_k, i_{k+1}, i_{k+2} are the corresponding currents and if the phase angle between the current and the voltage is θ, then

$$i_k = I \sin(\omega_0 t_k + \theta) \qquad (4.51)$$

$$i_{k+1} = I \sin(\omega_0 t_k + \omega_0 \Delta t + \theta) \qquad (4.52)$$

$$i_{k+2} = I \sin(\omega_0 t_k + 2\omega_0 \Delta t + \theta) \qquad (4.53)$$

By manipulating eqns. 4.48 to 4.53 and using appropriate trigonometric identities, the following relationships can be proved:

$$i_{k+1}^2 - i_k i_{k+2} = I^2 \sin^2(\omega_0 \Delta t) \qquad (4.54)$$

$$v_{k+1} i_{k+2} - v_{k+2} i_{k+1} = - VI \sin \theta \sin \omega_0 \Delta t \qquad (4.55)$$

$$2v_{k+1} i_{k+1} - v_{k+2} i_k - v_k i_{k+2} = 2VI \cos \theta \sin^2 \omega_0 \Delta t \qquad (4.56)$$

Now the resistive part (R_f) of the measured impedance is given by $R_f = (V/I) \cos \theta$ so that, by using eqns. 4.54 and 4.56, we obtain

$$R_f = \frac{2v_{k+1} i_{k+1} - v_{k+2} i_k - v_k i_{k+2}}{2(i_{k+1}^2 - i_k i_{k+2})} \qquad (4.57)$$

The reactive part (X_f) of the measured impedance is given by $X_f = (-V/I) \sin \theta$, and can likewise be readily determined from a three sample voltage and current measurand set. This is done by means of eqns. 4.54 and 4.55, which leads to:

$$X_f = \frac{v_{k+1} i_{k+2} - v_{k+2} i_{k+1}}{i_{k+1}^2 - i_k i_{k+2}} \sin \omega_0 \Delta t \qquad (4.58)$$

It should be noted that eqn. 4.57 shows that, as expected, R_f is independent of sampling rate and therefore is not affected by the system frequency. On the other hand, eqn. 4.58 shows that X_f is dependent on the constant $\sin \omega_0 \Delta t$; as expected, the measured reactance is therefore proportional to the system frequency.

4.6 An early relaying scheme [3, 4]

One practical distance protection scheme that uses the first- and second-derivative algorithm was jointly developed by Westinghouse Electric Corporation and Pennsylvania Power & Light Company. This is known as Prodar 70, and it was one of the earliest developments which provided the

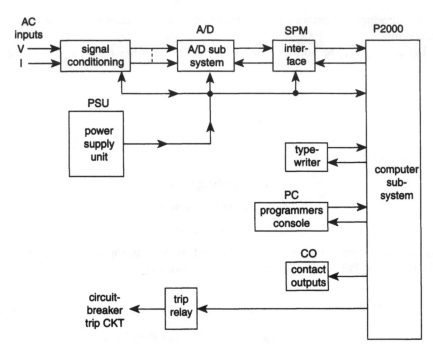

Figure 4.2   Schematic diagram for the basic components of the Prodar 70 protection system (quoted from Reference 3)

major functions of distance protection as well as instantaneous over-current protection and out-of-step blocking. Figure 4.2 shows a schematic diagram of the scheme, in which voltage and currents derived from the CTs and VTs are passed through a signal-conditioning unit and then to an analogue-to-digital subsystem where they are sampled and converted into digital form under the control of a data buffer (known as scratch pad memory or SPM). The digitised data are then transferred to the memory of the computer subsystem for processing. If, as a result of the processing, a fault is detected, a signal is initiated to trip the circuit breaker as well as for operating various event alarms. Some peripherals are included in the system. Examples of these are a KSR-typewriter and a programmed console to provide facilities for the logging of data as well as a convenient means of execution control, software generation and program checkout.

4.7 References

1 MANN, B.J., and MORRISON, I.F.: 'Digital calculation of impedance for trans-mission line protection', *IEEE Trans.* 1971, **PAS-90**, pp. 270–279
2 MANN, B.J., and MORRISON, I.F.: 'Relaying a three phase transmission line with a digital computer', *ibid*, 1971, pp. 742–750

3 GILCHRIST, G.B., ROCKEFELLER, G.D., and UDREN, E.A.: 'High speed distance relaying using a digital computer, part I—system description', *IEEE Trans.* 1972, **PAS-91**, pp. 1235–1243
4 ROCKEFELLER, G.D., and UDREN, E.A.: 'High-speed distance relaying using a digital computer, part II—test results', *ibid*, 1972, **PAS-91**, pp. 1244–1252
5 MAKINO, J., and MIKI, Y.: 'Study of operating principles and digital filters for protective relays with digital computers', IEEE PES Winter Power Meeting, New York, January 1975, 75CHO990-2 PWR, Paper C75, 197–9, pp. 1–8
6 GILBERT, J.G., and SHOVLIN, R.J.: 'High speed transmission line fault impedance calculation using a dedicated computer', *IEEE Trans.* 1975, **PAS-94**, pp. 872–883

Chapter 5

Fourier analysis and Walsh function based techniques

5.1 Introduction

In this Chapter we present digital relaying algorithms which are based on Fourier and Walsh analyses. From the point of view of this discussion, Fourier analysis includes Fourier series and Fourier transform-based methods, while Walsh analysis includes Walsh series only.

The basic assumption used in Fourier and Walsh series based methods is that the waveform that results from a fault condition (voltage and/or current) is assumed to be periodic within the interval, say, from t_0 to $t_0 + T$, where T is the period of the fundamental component. This assumption enables the waveform to be expanded by either Fourier or Walsh series. The fundamental component is then extracted and then used to calculate either the impedance to the fault or differential current quantities.

In the case of the Fourier transform method, no assumption as to the nature of faulted waveform is necessary. Both the voltage and current waveforms within the data window are transformed to the frequency domain. These transformed quantities are then used to calculate the apparent impedance to the fault.

5.2 Fourier-analysis-based algorithms

It was shown in Section 2.6.1 that any function of time $(f(t))$ can be represented by a Fourier series and each coefficient of the series can be found according to the formula given in eqns. 2.60–2.62.

Voltage and current waveforms are of course functions of time and they can be consequently expanded using the Fourier series. If we take, for example, a voltage waveform $v(t)$, then

$$v(t) = \frac{a_0}{2} + \sum_{n=1}^{\infty} a_n \cos n\omega_0 t + \sum_{n=1}^{\infty} b_n \sin n\omega_0 t \qquad (5.1)$$

and from eqns. 2.61 and 2.62

$$a_n = \frac{2}{T} \int_{t_0}^{t_0+T} v(t) \cos n\omega_0 t \, dt, \; n = 0, 1, \ldots \qquad (5.2)$$

$$b_n = \frac{2}{T} \int_{t_0}^{t_0+T} v(t) \sin n\omega_0 t \, dt, \quad n = 1, 2, \ldots \tag{5.3}$$

where ω_0 is the angular frequency of the fundamental component and T is its period.

Eqns. 5.2 and 5.3 show that the fundamental component of a voltage and/or current waveform can be extracted from the corresponding faulted waveform simply by setting $n = 1$.

5.2.1 The full cycle window algorithm [1–3]

Basic approach
The basic approach used in this algorithm is to extract the fundamental component of a waveform by correlating one cycle of the faulted waveform with stored reference sine and cosine waves, as outlined in the previous Section.

Derivation of the algorithm
Let V_x and V_y be the real and imaginary parts, respectively, of the phasor that represents the fundamental component of the faulted voltage waveform $v(t)$. If the time under consideration is t_0, then V_x can be found from eqn. 5.2 such that

$$V_x = a_1 = \frac{2}{T} \int_{t_0}^{t_0+T} v(t) \cos \omega_0 t \, dt \tag{5.4}$$

Now let N be the number of samples per cycle of the fundamental component, Δt the sampling time interval, $t_j = j\Delta t$ the time of the jth sample and $T = N\Delta t$ the period of the fundamental component.

The integral that appears in eqn. 5.4 can then be evaluated using the rectangular method, which leads to

$$V_x \approx \frac{2}{N\Delta t} [v(t_0) \cos \omega_0 t_0 + v(t_1) \cos \omega_0 t_1 + \ldots v(t_j) \cos \omega_0 t_j$$

$$+ \cdots + v(t_{N-1}) \cos \omega_0 t_{N-1} + v(t_N) \cos \omega_0 t_N] \Delta t$$

$$V_x \approx \frac{2}{N} \sum_{j=0}^{N} v_j \cos \left(\frac{2\pi j}{N} \right) = \frac{2}{N} \sum_{j=0}^{N} W_{x,j} v_j \tag{5.5}$$

where $v_j = v(t_j)$ is the jth sample of the voltage waveform, and $W_{x,j}$ is the weighting factor of the jth sample used to calculate V_x, which in turn is given by

$$W_{x,j} = \cos \omega_0 t_j = \cos \frac{2\pi}{T} j\Delta t$$

or, in the alternative form of

$$W_{x,j} = \cos(2\pi j / N), \quad j = 0, 1, \ldots N \tag{5.6}$$

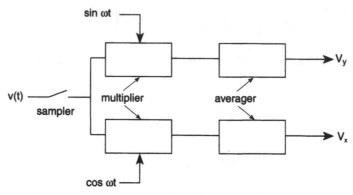

Figure 5.1   Schematic representation of the full-cycle window Fourier algorithm

Similarly, it is possible to approximate V_y in terms of discrete waveform samples and their weighting factors as follows:

$$V_y \approx \frac{2}{N} \sum_{j=0}^{N} v_j \sin\left(\frac{2\pi j}{N}\right) = \frac{2}{N} \sum_{j=0}^{N} W_{y,j} v_j \qquad (5.7)$$

where $W_{y,j}$ is the weighting factor of the jth sample used to calculate V_y, as defined in eqn. 5.8

$$W_{y,j} = \sin(2\pi j/N), \qquad j = 0, 1, \ldots N \qquad (5.8)$$

The algorithm described by eqns. 5.5 to 5.8 can also be used to calculate the real and imaginary parts I_x and I_y of the fundamental current component from current waveform samples, and Figure 5.1 shows a schematic representation of the algorithm.

The weighting factors defined by eqns. 5.6 and 5.8 can be made to have values of 0, ± 1, $\pm 1/2$, and $\pm \sqrt{3}/2$, which in turn are suitable for computer application, if the sampling rate is suitably chosen. For example, if $N = 12$, then the values of $W_{x,j}$ and $W_{y,j}$ are as given in Table 5.1. This technique reduces the

Table 5.1   Weighting factors for $N = 12$ samples per fundamental cycle

$W_{x,j}$	j	$W_{y,j}$	j
1	0, 12	0	0, 6, 12
$\sqrt{3}/2$	1, 11	1/2	1, 5
1/2	2, 10	$\sqrt{3}/2$	2, 4
0	3, 9	1	3
$-\sqrt{3}/2$	4, 8	$-1/2$	7, 11
$-1/2$	5, 7	$-\sqrt{3}/2$	8, 10
-1	6	-1	9

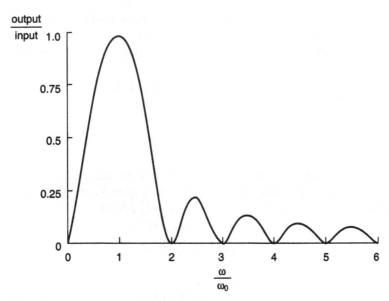

Figure 5.2 Frequency response of the full-cycle window Fourier algorithm

amount of computation involved in applying the method in the real-time mode of operation. Lower sampling rates, such as four or eight per cycle, can be used where the application permits, but for many transmission-line applications $N = 12$ is most appropriate.

When the weighting factors of Table 5.1 are used, eqns. 5.5 and 5.7 take the following simplified form:

$$V_x = \frac{1}{6}\left[v_0 + v_{12} + \frac{1}{2}(v_2 - v_5 + v_{10} - v_7) + \frac{\sqrt{3}}{2}(v_1 - v_4 + v_{11} - v_8) \right] \quad (5.9)$$

$$V_y = \frac{1}{6}\left[\left(v_3 - v_9 + \frac{1}{2}(v_1 - v_7 + v_5 - v_{11}) \right) + \frac{\sqrt{3}}{2}(v_2 - v_8 + v_4 - v_{10}) \right] \quad (5.10)$$

Figure 5.2 represents the frequency response of a full-cycle window algorithm from which DC and/or harmonic components are completely filtered out. This is a useful and important property because it renders subsequent calculations, e.g. circuit impedances, completely immune to error due to discrete harmonic distortion of the signal measurands.

5.2.2 Fractional-cycle window algorithms

5.2.2.1 Half-cycle window algorithm [4]
This algorithm is based on correlating the fault waveforms with sine and cosine functions having a frequency equal to that of the fundamental component of the waveform. It is therefore basically the same as that of the full-cycle window, except that it uses information corresponding to only one half cycle.

Let $V_{x, 1/2}$ and $V_{y, 1/2}$ be the real and imaginary parts of the phasor that represents the fundamental component derived from a half-cycle window. Thus, by applying eqns. 5.2 and 5.3 to a half-cycle window, we obtain

$$V_{x, 1/2} = \frac{2}{(T/2)} \int_{t_0}^{t_0 + T/2} v(t) \cos \omega_0 t \, dt \qquad (5.11)$$

$$V_{y, 1/2} = \frac{2}{(T/2)} \int_{t_0}^{t_0 + T/2} v(t) \sin \omega_0 t \, dt \qquad (5.12)$$

Following the same procedure as that followed in Section 5.2.1, it is possible to express $V_{x, 1/2}$ and $V_{y, 1/2}$ in terms of waveform samples S_j, and their corresponding weighting factor $W_{x, j}$ and $W_{y, j}$, $(j = 1, 2, \ldots, N/2)$ such that

$$V_{x, 1/2} = \frac{4}{N} \sum_{j=1}^{N/2} W_{x, j} v_j \qquad (5.13)$$

$$V_{y, 1/2} = \frac{4}{N} \sum_{j=1}^{N/2} W_{y, j} v_j \qquad (5.14)$$

If $N = 12$, then the waveform coefficients can be written in the simplified form

$$V_{x, 1/2} = \frac{1}{3} \left[v_0 - v_6 + 1/2(v_2 - v_5) + \frac{\sqrt{3}}{2} (v_1 - v_4) \right] \qquad (5.15)$$

$$V_{y, 1/2} = \frac{1}{3} \left[v_3 + \frac{1}{2} (v_1 + v_5) + \frac{\sqrt{3}}{2} (v_2 + v_4) \right] \qquad (5.16)$$

By comparing eqns. 5.5 and 5.7 with eqns. 5.13 and 5.14, or eqns. 5.9 and 5.10 with 5.15 and 5.16, it can be seen that the half-cycle window algorithm uses only half the number of samples of those used in the full-cycle algorithm. However, although this algorithm appears to be faster than that of the alternative it has the disadvantage of introducing error, specifically due to any aperiodic component and/or even harmonics. This can be seen in Figure 5.3 which represents the frequency response of the half-cycle-window algorithm. The useful property of a rejection of odd harmonics is however retained.

5.2.2.2 Sub-cycle window algorithm [5, 6]

In an attempt to reduce the error introduced into half-cycle windows in the presence of aperiodic components, the sub-cycle window algorithm correlates faulted waveforms with sine/cosine functions having a period equal to the data window length.

In this algorithm, the current and voltage waveforms are assumed to have the following mathematical form.

$$i(t) = I_0 \, e^{-t/\tau_0} + I_1 \cos(\omega_0 t - \theta_{f1}) + I_p \, e^{-t/\tau_p} \cos(\omega_p t - \gamma) \qquad (5.17)$$

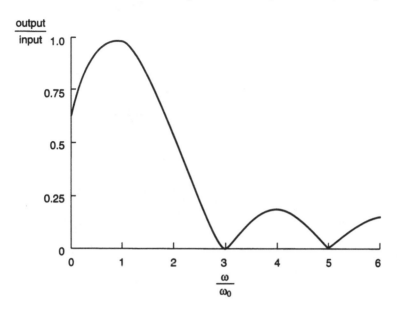

Figure 5.3   Frequency response of the half-cycle window Fourier algorithm

$$v(t) = V_0 \, e^{-t/\tau_0} + V_1 \, \cos(\omega_0 t - \theta_{V1}) + V_p \, e^{-t/\tau_p} \, \cos(\omega_p t - \beta) \qquad (5.18)$$

where

I_0, V_0 are the amplitudes of any decaying DC current and voltage components
I_1, V_1 are the amplitudes of the fundamental components of current and voltage
I_p, V_p are the amplitudes of any assumed decaying current and voltage oscillations
τ_0, τ_p are the decaying time constants for any DC and high frequency oscillations and
ω_0, ω_p are the frequencies of the fundamental and any high-frequency oscillations.

Since the data processed falls within a window (of length T_w) it is convenient to take the middle of the window as a reference. This can be achieved by shifting the time scale t to a new variable t', as shown in Figure 5.4, such that

$$t' = t - (t_1 + T_w/2) = t - t_1' \qquad (5.19)$$

where

$$t_1' = t_1 + T_w/2$$

If V_x' and V_y' are the quantities resulting from the correlation of the shifted

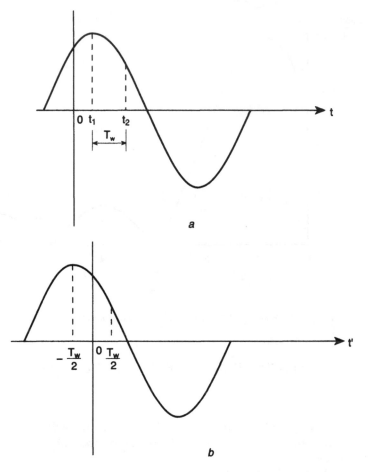

*Figure 5.4 Shifting of the time scale t to the variable t' where $t' = t - (t_1 + T_w/2)$.
(a) = original waveform, (b) = shifted version*

* purely sinusoidal waveform is assumed

waveform $v(t')$ with cosine and sine functions having a period equal to the time span of data window T_w, then

$$V'_x = \int_{-T_w/2}^{T_w/2} v(t') \cos \omega_w t' \, dt' \qquad (5.20)$$

$$V'_y = \int_{-T_w/2}^{T_w/2} v(t') \sin \omega_w t' \, dt' \qquad (5.21)$$

where

$$\omega_w = \frac{2\pi}{T_w}$$

is the frequency of correlating sine/cosine function.

The amplitude V_1 and phase angle θ_{V1} of the fundamental component of $v(t)$ can be determined from V'_x, V'_y by using the following equations:

$$V_1 = \frac{1}{T_w} \sqrt{(kV'_x)^2 + (pV'_y)^2}$$

$$\theta_{V1} = \tan^{-1}(pV'_y/kV'_x) \tag{5.22}$$

The factors k and p depend on data window length T_w, and they can be determined as follows:

$$k = \frac{\pi(1 - r^2)}{r \sin \pi r}$$

$$p = \frac{-\pi(1 - r^2)}{r \sin \pi r} \tag{5.23}$$

where $r = \omega_w/\omega_0$.

Although this algorithm reduces the error due to any aperiodic components, it causes the error due to high-frequency oscillations to increase. From a practical point of view, this algorithm is therefore most suitable for applications where oscillatory components in the currents and voltages are relatively small, e.g. in distribution system applications.

5.2.3 Fourier-transform-based algorithm [7]

Basic approach
This technique does not impose any assumption as to the nature of the relaying waveform within any chosen window of information analysed. The basic approach involves applying the Fourier transform to a data window within the voltage and current waveforms. The impedance is then calculated from the voltage and current transforms. By progressively advancing the window, it is possible to observe the impedance variation with time after a fault, which in turn is used to determine whether there is a fault within the protected zone.

Outline of the algorithm
Consider the assumed current and voltage waveforms shown in Figure 5.5, where the window is shown to be from T_1 to T_2. Since we are interested in finding the Fourier transform of this window of information, we shall concentrate our attention on the voltage and current quantities therein. Let us assume that the voltage waveform $v(t)$ is made up of three components, such that

$$v(t) = v_1(t) + v_2(t) + v_3(t) \tag{5.24}$$

with

$$v_1(t) = \begin{cases} v(t) & t < T_1 \\ 0 & t > T_1 \end{cases} \tag{5.25}$$

$$v_2(t) = \begin{cases} v(t) & T_1 \leqslant t \leqslant T_2 \\ 0 & T_1 > t > T_2 \end{cases} \tag{5.26}$$

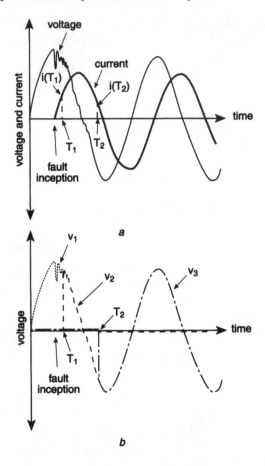

Figure 5.5   (a) Faulted voltage and current waveform (b) Decomposition of voltage waveform v into three components v_1, v_2, v_3

$$v_3(t) = \begin{cases} v(t) & t > T_2 \\ 0 & t < T_2 \end{cases} \tag{5.27}$$

According to the principles of circuit theory, each of the voltages v_1, v_2 and v_3 would produce a corresponding current according to the following equation:

$$v_k(t) = Z(p)i_k(t), \qquad k = 1, 2, 3 \tag{5.28}$$

where $Z(p)$ is the operational value of the circuit impedance.

By the principle of superposition, the three components add to give the total current $i(t)$, such that

$$i(t) = i_1(t) + i_2(t) + i_3(t) \tag{5.29}$$

Now the Fourier transform of the window of voltage and current, $v_2(t)$ and $i_2(t)$, can be calculated using eqn. 2.73 to obtain

$$v_2(j\omega) = \int_{-\infty}^{\infty} v_2(t)e^{-j\omega t}dt \qquad (5.30)$$

The limits defined in eqn. 5.26 allow the Fourier transform of the given window to be reduced:

$$v_2(j\omega) = \int_{T_1}^{T_2} v(t)e^{-j\omega t}dt \qquad (5.31)$$

The Fourier transform of the corresponding current $i_2(t)$ is given by

$$i_2(j\omega) = \int_{-\infty}^{\infty} i_2(t)e^{-j\omega t}dt \qquad (5.32)$$

It will be noted that $v_2(j\omega)$ can be easily evaluated using numerical methods (such as the trapezoidal technique), because $v_2(t)$ is readily available. However, this is not the case for $i_2(j\omega)$ because $i_2(t)$ is not readily available.

Before proceeding to a derivation of $i_2(j\omega)$, it is useful to note the following points:

(i) According to the principles of circuit theory [8], when the voltage collapses, the current in the circuit does not cease immediately. Therefore, if, for example, voltage v_1 collapses at $t = T_c$, the current $i_1(t)$ which flows for time $t > T_c$ is approximately

$$i_1(t) = i(T_c)e^{-(t-T_c)/\tau} \qquad t > T_c \qquad (5.33)$$

where τ is the time constant of the circuit.

(ii) The current that would flow within the data window consists of two components, the first being due to the voltage component $v_1(t)$ (as explained in (i) above), and the second due to the voltage component $v_2(t)$.

(iii) By the same reasoning as given in (i) above, the current $i_2(t)$ is not confined to the window, i.e. it does not cease at time T_2, but rather continues to flow thereafter, despite the collapse of voltage v_2 at T_2.

Now if the total current within the window is $i(t)$, then

$$i_2(t) = i(t) - i_1(t) \qquad T_1 \leqslant t \leqslant T_2 \qquad (5.34)$$

But, according to eqn. 5.33, $i_1(t)$ at $t > T_1$ would be given by substituting T_c by T_1

$$i_1(t) = i(T_1)e^{-(t-T_1)/\tau} \qquad T_1 \leqslant t \leqslant T_2 \qquad (5.35)$$

By substituting this equation into eqn. 5.34, we obtain

$$i_2(t) = i(t) - i(T_1)e^{-(t-T_1)/\tau} + f_{h1}(t) \qquad T_1 \leqslant t \leqslant T_2 \qquad (5.36)$$

It must be emphasised that this equation is valid only during the window bounded by $T_1 \leqslant t \leqslant T_2$.

At $t = T_2$ the voltage component $v_2(t)$ collapses and therefore $i_2(t)$ continues to flow according to eqn. 5.33 (see remarks (i) and (iii)):

$$i_2(t) = i_2(T_2)e^{-(t-T_2)/\tau} + f_{h2}(t) \qquad\qquad t > T_2 \qquad\qquad (5.37)$$

where $i_2(T_2)$ is the value of the current $i_2(t)$ at $t = T_2$, and can be found from eqn. 5.36 by substituting t by T_2.

$$i_2(T_2) = i(T_2) - i(T_1)e^{-(T_2-T_1)/\tau} + f_{h1}(T_2) \qquad\qquad (5.38)$$

By combining eqns. 5.37 and 5.38, we obtain the current which flows in response to the application of the window bounded by times T_1, T_2:

$$i_2(t) = [i(T_2) - i(T_1)e^{-(T_2-T_1)/\tau} + f_{h1}(T_2)]e^{-(t-T_2)} + f_{h2}(t) \qquad t > T_2 \quad (5.39)$$

The components $f_{h1}(t)$ and $f_{h2}(t)$ are added to take into account any high-frequency components caused by travelling waves. However, practically, for lines of length not more than 250 km, these high-frequency components can be neglected because their magnitudes are small compared with the components at or near power frequency. Therefore, by substituting $i_2(t)$ described by eqns. 5.36 and 5.39 into eqn. 5.32, and neglecting $f_{h1}(t)$ and $f_{h2}(t)$, we obtain:

$$i_2(j\omega) = \int_{T_1}^{\infty} i_2(t)e^{-j\omega t}dt = \int_{T_1}^{T_2} i_2(t)e^{-j\omega t}dt + \int_{T_2}^{\infty} i_2(t)e^{-j\omega t}dt$$

or

$$i_2(j\omega) = \int_{T_1}^{T_2} \left[i(t) - i(T_1)e^{-(t-T_1)/\tau} \right] e^{-j\omega t}dt$$

$$+ \int_{T_2}^{\infty} [i(T_2) - i(T_1)e^{-(T_2-T_1)/\tau}]e^{-(t-T_2)/\tau}e^{-j\omega t}dt$$

which results in

$$i_2(j\omega) = \bar{i}(j\omega) + \frac{\tau}{1+j\omega\tau} [i(T_2)e^{-j\omega T_2} - i(T_1)e^{-j\omega T_1}] \qquad\qquad (5.40)$$

where

$$\bar{i}(j\omega) = \int_{T_1}^{T_2} i(t)e^{-j\omega t}dt \qquad\qquad (5.41)$$

Computational aspects

Since voltage and current waveforms are presented to the algorithm in the form of samples, the integration terms that appear in $v_2(j\omega)$ and $i_2(j\omega)$ should therefore be evaluated numerically.

Let Δt be the sampling interval. Then

$$\begin{aligned} T_1 &= n\Delta t \\ T_w &= N\Delta t \\ T_2 &= T_1 + T_w \\ &= (N+n)\Delta t \end{aligned} \qquad\qquad (5.42)$$

where n is an integer which defines the position of the data window, and N is the number of sampling intervals within the data window of length T_w.

By substituting T_1, T_w and T_2 (defined in eqn. 5.42) into eqn. 5.31, we obtain

$$v_{2n}(j\omega) = \int_{n\Delta t}^{(n+N)\Delta t} v(t)e^{-j\omega t}dt$$

where $v_{2n}(j\omega)$ is the transform of $v_2(t)$ for a window starting at $T_1 = n\Delta t$. When the trapezoidal rule is applied to this equation, we obtain

$$v_{2n}(j\omega) \approx \left\{ \frac{1}{2}\, v(n\Delta t)e^{-jn\omega\Delta t} + v[(n+1)\Delta t]e^{-j(n+1)\omega\Delta t} + \cdots \right.$$

$$\left. + v[(n+N-1)\Delta t]e^{-j(n+N-1)\omega\Delta t} + \frac{1}{2}\, v[(n+N)\Delta t]e^{-j(n+N)\omega\Delta t} \right\}\Delta t$$

or

$$v_{2n}(j\omega) = \frac{\Delta t}{2}\left[W_n v_n + W_{n+N}v_{n+N} + \sum_{m=1}^{N-1} 2W_{n+m}v_{n+m} \right] \qquad (5.43)$$

where v_n is the nth voltage sample, and W_{n+m} is the weighting factor of the $(n+m)$th voltage sample.

When the window is moved by one sample, the transform of the voltage $v_2(t)$ within the new window, i.e. $v_{2(n+1)}(j\omega)$, can be found recursively by simply dropping the oldest sample v_n and adding the newest sample v_{n+N+1}, such that

$$v_{2, (n+1)}(j\omega) = v_{2n}(j\omega) + Q_n \qquad (5.44)$$

where

$$Q_n = -\frac{\Delta t}{2}\left[-W_n v_n - W_{n+1}v_{n+1} + W_{n+N}v_{n+N} + W_{n+N+1}v_{n+N+1} \right]$$

This equation can be verified mathematically by following the foregoing procedure and substituting for $T_1 = (n+1)\Delta t$ and $T_2 = (n+N+1)\Delta t$. The resulting equation is then reduced to the form described by eqn. 5.44. Thus, to update the voltage transform as the window advances requires only the addition of the sampled function Q_n to the integral approximation taken over the period immediately before the one in question.

Computational efficiency can be improved if the extraction frequency ω is so chosen so that the period of Fourier kernel terms ($e^{-j\omega n\Delta t}$) corresponds to an integer multiple k of the sampling interval, i.e., $2\pi/\omega = k\Delta t$. If this is done, the weighting coefficients W_{n+m} can be pre-evaluated and used in a cyclic process as the window moves, thus obviating the necessity for an online evaluation of the

Fourier kernels. The previous method can be used to evaluate the current transform $i_2(j\omega)$ defined by eqns. 5.40 and 5.41, which in turn results in

$$i_{2n}(j\omega) = \frac{\Delta t}{2}\left[W_n i_n + W_{n+N} i_{n+N} + \sum_{m=1}^{N-1} W_{n+m} i_{n+m} \right] + \frac{u_n}{1+j\omega T} \qquad (5.45)$$

where $u_n = W_{n+N} i_{n+N} - W_n i_n$, $i_n = n$th current sample, $i_{2n}(j\omega) =$ the transform of $i_2(t)$ relating to the window which starts at $T = n\Delta t$.

5.3 Walsh-function-based algorithms

Digital relaying using Walsh functions was originally developed to reduce the burden on computers by using only add and shift operations in fixed point format, thereby avoiding the multiply, divide, square and square-root operations. The computational simplification is made possible by the fact that Walsh functions assume values of only ± 1.

5.3.1 Basic principles

The basic principles of digital protection using Walsh function techniques consists of expanding the faulted waveforms of voltage and current by using Walsh series, from which Walsh coefficients $\{W_k, k = 1, 2, \ldots\}$ are determined. These functions are then used either indirectly, through the Fourier–Walsh relation, or directly to determine the amplitude and phase angle of the fundamental components of the voltage and current waveforms, which are then used to determine the impedance to the fault.

5.3.2 Development of basic algorithm

It will be seen from Sections 2.6 and 2.7 that any assumed periodic function can be expanded using either the Fourier series or the Walsh series. Therefore, if for example a faulted voltage waveform $v(t)$ is assumed to be periodic in the interval from t_1 to $t_1 + T$ (assuming $t_1 = 0$), then, from eqns. 2.83 and 2.84 we obtain

$$v(t) = F_0 + \sum_{n=1}^{\infty} (\sqrt{2}F_{2n-1} \sin n\omega_0 t + \sqrt{2}F_{2n} \cos n\omega_0 t) \qquad (5.46)$$

with

$$F_0 = \frac{1}{T}\int_0^T v(t)dt$$

$$F_{2n-1} = \frac{\sqrt{2}}{T}\int_0^T v(t) \sin n\omega_0 t \, dt \qquad (5.47)$$

$$F_{2n} = \frac{\sqrt{2}}{T} \int_0^T v(t) \cos n\omega_0 t \, dt$$

where F_0 is the DC component and F_{2n}, F_{2n+1} are the RMS values of the real and imaginary components of the nth harmonic, respectively. As indicated previously, it is the Fourier series component values at fundamental (or power) frequency that are used to determine the measured impedance. However, it is relatively easy to determine these by first evaluating the Walsh component values. In this respect, it will be recalled that by using the Walsh series, $v(t)$ can be approximated using eqns. 2.85 to obtain

$$v(t) = \sum_{k=0}^{\infty} W_k \, \mathrm{Wal}\left(k, \frac{t}{T}\right) \tag{5.48}$$

with the kth Walsh coefficient (W_k) given by

$$W_k = \frac{1}{T} \int_0^T v(t) \, \mathrm{Wal}\left(k, \frac{t}{T}\right) dt \qquad k = 0, 1, 2, \ldots \tag{5.49}$$

Fourier and Walsh coefficients are interrelated by matrix A, as described by eqn. 2.92 which is rewritten below:

$$F = A^T W \tag{5.50}$$

The Fourier to Walsh transformation matrix A and its transpose consist of fixed constant values (see eqn. 2.90) and, given a determination of Walsh components using eqn. 5.49, it is relatively straightforward to determine the fundamental or where required harmonic Fourier series components using eqn. 5.50 above.

5.3.3 Algorithm for Walsh function determination [9]

It is possible to reduce the amount of real-time computation by employing a fast algorithm, which is used to calculate the Walsh function coefficients W_k for any arbitrary waveform. Consider a voltage waveform that is available in the form of samples at a rate of N samples per cycle. In this case, the sampling interval Δt is related to the time over which the signal is periodic (T) and the number of samples therein (N) by $T = N\Delta t$. In this case $t_j = j\Delta t$, $j = 0, 1, 2, \ldots, N$ and $v(t_j) = v(j\Delta t)$. Then

$$\mathrm{Wal}(k, t_j') = \mathrm{Wal}\left(k, \frac{j\Delta t}{T}\right) = \mathrm{Wal}(k, j\Delta t'),$$

where

$$\Delta t' = \frac{\Delta t}{T}$$

When this is done, the Walsh coefficients W_k can be calculated according to eqn. 5.49, such that

$$W_k = \frac{1}{T} \int_0^T v(t) \cdot \text{Wal}(k, t') dt$$

The integral on the right hand side of the above equation can usefully be evaluated using the trapezoidal rule, which results in

$$W_k = \frac{1}{N\Delta t} \left[\frac{1}{2} v(t_0) \cdot \text{Wal}(k, 0) + v(t_1) \cdot \text{Wal}(k, \Delta t') + \cdots + v(t_j) \cdot \text{Wal}(k, j\Delta t') \right.$$

$$\left. + v(t_{N-1}) \cdot \text{Wal}(k, (N-1)\Delta t') + \frac{1}{2} v(t_N) \cdot \text{Wal}(k, N\Delta t') \right] \Delta t \qquad (5.51)$$

Now let $v(t_j) = v_j$, $w_j(k) = \text{Wal}(k, j\Delta t')$ and $f_j = v(t_j) \cdot \text{Wal}(k, j\Delta t') = v_j \cdot w_j(k)$, so that by substituting into eqn. 5.51 we obtain

$$W_k = \frac{1}{N} \left[\frac{1}{2} f_0 + f_1 + \ldots f_j + \ldots f_{N-1} + \frac{1}{2} f_N \right] \qquad (5.52)$$

Walsh coefficients W_k ($k = 1, 2, ..$) are evaluated as each new sample of $v(t)$ is taken. This can be done by dropping off v_0 (the oldest value) and adding or subtracting v_{N+1} (the newest value). It is possible to imagine the waveform $v(t)$ as passing from right to left through a window which contains the Walsh components. Figure 5.6 shows the effect of movement from right to left of a sampled sinusoidal waveform $v(t)$ through a window which contains the Walsh function $\text{Wal}(1, t')$. The effect of this movement on resulting $f(j\Delta t)$ is also shown.

Generally, if the voltage waveform moves an equivalent of $s\Delta t$ with respect to the initial position of sample v_j, then the sample which occupies the position of v_j will be v_{j+s}. This simply means that moving $v(t)$ s steps leftwards is equivalent to a time displacement of $s\Delta t$. This in turn results in

$$f_{j+s} = v_{j+s} \cdot w_j(k) \qquad s = 0, 1, 2, \ldots \qquad (5.53)$$

Therefore, by moving the waveform $v(t)$ by $s \cdot \Delta t$, the Walsh coefficient $W_k(s)$ is given as

$$W_k(s) = \frac{1}{N} \left[\frac{1}{2} f_s + f_{1+s} + \ldots f_{j+s} + \ldots f_{N-1+s} + \frac{1}{2} f_{N+s} \right] \qquad (5.54)$$

If for example $N = 8$, then either with the help of Figure 2.6 or by using the discrete representation of the Walsh function given in Section 2.7.3, it can be seen that

$$W_0(s) = \frac{1}{8} \left[\frac{1}{2} v_s + v_{s+1} + \ldots v_{7+s} + \frac{1}{2} v_{8+s} \right] \qquad (5.55)$$

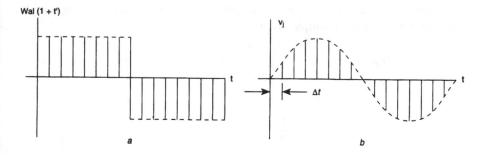

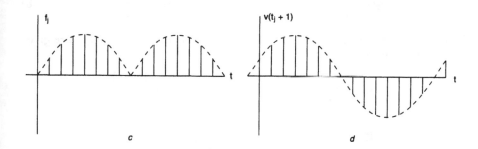

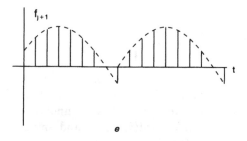

Figure 5.6 *The mechanism of movement from right to left of a waveform through a window that contains* Wal(1, *t*)

(a) Walsh function window which contains Wal(1, t')
(b) Sampled sinusoidal waveform $v_j = v(t_j) = v(j\Delta t)$
(c) Resulting function $f(j\Delta t) = v(t_j) \cdot \text{Wal}(1, t')$
(d) The function $v(t_{j+1}) = v((j+1)\Delta t)$ resulting from $v(t_j)$ after being moved to the left by one sampling interval
(e) The resulting $f_{j+1} = f((j+1)\Delta t) = v((j+1)\Delta t) \cdot \text{Wal}(1, t')$

Therefore $W_0(s+1)$ can be found recursively by dropping off the oldest sample and adding the newest sample v_{9+s} to obtain

$$W_0(s+1) = W_0(s) + \frac{1}{8}\left[-\frac{1}{2}v_s - \frac{1}{2}v_{1+s} + \frac{1}{2}v_{8+s} + \frac{1}{2}v_{9+s} \right] \qquad (5.56)$$

Similarly, W_1 can be found recursively as

$$W_1(s+1) = W_1(s) + \frac{1}{8}\left[-\frac{1}{2}v_s - \frac{1}{2}v_{1+s} + v_{4+s} + v_{5+s} - \frac{1}{2}v_{8+s} - \frac{1}{2}v_{9+s} \right]$$

$$s = 0, 1, 2, \ldots \qquad (5.57)$$

From a computational point of view, not much is gained from calculating W_2 recursively since it can be calculated directly as

$$W_2(s) = \frac{1}{8}\left[\frac{1}{2}v_s + v_{1+s} - v_{3+s} - v_{4+s} - v_{5+s} + v_{7+s} + \frac{1}{2}v_{8+s} \right] \qquad (5.58)$$

The remaining coefficients can be found in exactly the same way and are given in eqn. 5.59 below.

$$W_3(s) = \frac{1}{8}\left[\frac{1}{2}v_s + v_{1+s} - v_{3+s} + v_{5+s} - v_{7+s} - \frac{1}{2}v_{8+s} \right]$$

$$W_4(s) = \frac{1}{8}\left[\frac{1}{2}v_s - v_{2+s} + v_{4+s} + \frac{1}{2}v_{8+s} \right]$$

$$W_5(s) = \frac{1}{8}\left[\frac{1}{2}v_s - v_{2+s} + v_{6+s} - \frac{1}{2}v_{8+s} \right] \qquad (5.59)$$

$$W_6(s) = \frac{1}{8}\left[\frac{1}{2}v_s - v_{4+s} + \frac{1}{2}v_{8+s} \right]$$

$$W_7(s) = \frac{1}{8}\left[\frac{1}{2}v_s - \frac{1}{2}v_{8+s} \right]$$

It will be noted that, except for the end-points of the interval $(0, T)$, the application of the trapezoidal rule leads to the cancellation of the sample value where a jump occurs in the appropriate Walsh function. This is because the area of the function $f(t_j) = v(t_j) \cdot \text{Wal}(k, t_j')$ around any such jump has equal parts but with opposite sign (see Figure 5.6e).

5.3.4 Estimation of the amplitude and phase angle of fundamental components

5.3.4.1 Indirect method
This method involves the determination of the amplitude and phase angle of the fundamental component of a faulted voltage or current waveform using Walsh coefficients and the Walsh–Fourier transformation matrix A. Let $W_{k, v}$, $k =$

0, 1, 2, . . . be the Walsh coefficients of a faulted voltage waveform, determined as discussed in the previous subsection. Now, using these coefficients, it is possible to estimate the Fourier coefficients $F_{1, v}$ and $F_{2, v}$, which are related to the fundamental components. If we consider a 10×10 A-matrix, then, by eqn. 5.50, $F_{1, v}$ and $F_{2, v}$ can be expressed in terms of $W_{1, v}$ as follows:

$$F_{1, v} = 0.9 \ W_{1, v} - 0.373 \ W_{5, v} - 0.074 \ W_{9, v}$$
$$F_{2, v} = 0.9 \ W_{2, v} + 0.373 \ W_{6, v} - 0.074 \ W_{10, v} \tag{5.60}$$

It is then relatively easy to calculate the amplitude and phase angle of the fundamental Fourier series component from eqn. 5.61.

$$V_1 = \sqrt{(\sqrt{2} F_{1, v})^2 + (\sqrt{2} F_{2, v})^2}$$
$$= \sqrt{2} \ \sqrt{F_{1, v}^2 + F_{2, v}^2} \tag{5.61}$$

and

$$\theta_{v1} = \tan^{-1}(F_{2, v} / F_{1, v})$$

The amplitude and phase angle of the fundamental component of the current waveform can likewise be estimated in exactly the same way.

5.3.4.2 Direct method [9]

This method was developed principally to avoid the square root and squaring operations necessary when using previous methods. It also involves using Walsh coefficients directly (i.e. without going into a determination of the Fourier coefficient) to estimate the fundamental amplitude and its phase angle; this is based on two theorems.

The first theorem states that 'the maximum value of a sinusoid and the time of its occurrence can be estimated in terms of the first and second Walsh coefficients for the sinusoid'.

Assume $g_1(t)$ is a sinusoidal waveform with interval T, which is divided into 16 sub-intervals such that $\Delta t = T/16$. If $g_1(t)$ is expressed within T in terms of Walsh functions, such that

$$g_1(t) = \sum_{k=1}^{10} W_k \ \mathrm{Wal}\left(k, \frac{t}{T}\right) \tag{5.62}$$

then

(i) to an accuracy of $(+4.28\%$ to $-4.8\%)$ the maximum value of $g_1(t)$

(G_{max}) is given as $G_{max} = 1.0822(|W_1| + |W_2|) + 0.414\|W_1| - |W_2\|$ (5.63)

(ii) the maximum value of $g_1(t)$ occurs in the mth sub-interval, within which the following formulas are satisfied:

$$\mathrm{Wal}(1, \ m\Delta t) = S(W_1)$$
$$\mathrm{Wal}(2, \ m\Delta t) = S(W_2)$$
$$\mathrm{Wal}(3, \ m\Delta t) = S(W_2) \ . \ S(|W_2| - |W_1|) \tag{5.64}$$
$$\mathrm{Wal}(10, \ m\Delta t) = -S(W_2)$$

where $S(x) = \text{sign}(x)$ is a function which is positive when x is positive, and negative when x is negative. For $x = 0$ both plus and minus signs are to be used, and the maximum is not unique.

The second theorem states that the accuracy of estimating the amplitude of a sinusoid and its time of occurrence can both be improved if all 16 Walsh functions with unit interval are taken into account. Therefore, according to this second theorem, if $g_1(t)$ is a given sinusoidal function with periodic time T, which is divided into 16 sub-intervals, and if $g_1(t)$ is expanded in terms of the 16 Walsh functions defined by eqn. 5.65,

$$g_1(t) = \sum_{k=1}^{15} W_k \, \text{Wal}\left(k, \frac{t}{T}\right) \tag{5.65}$$

then

(i)   to an accuracy of $\pm 2.6\%$, the maximum value of $g_1(t)$, G_{max}, can be expressed in terms of W_1 and W_2 by the following formulas:

$$G_{max} = \begin{cases} (\alpha - \gamma)|\Delta| + (1 + \beta) & \text{if } (\alpha - \gamma)|\Delta| \leqslant \beta \sum \\ \\ (\alpha + \gamma)|\Delta| + (1 - \beta) & \text{if } (\alpha - \gamma)|\Delta| > \beta \sum \end{cases}$$

(ii) the maximum occurs in the mth sub-interval which satisfies the following:

$$\text{Wal}(1, \, m\Delta t) = S(W_1)$$

$$\text{Wal}(2, \, m\Delta t) = S(W_2)$$

$$\text{Wal}(6, \, m\Delta t) = S(W_2) \cdot S(\Delta)$$

$$\text{Wal}(10, \, m\Delta t') = \begin{cases} -S(W_2) & \text{if } (\alpha - \gamma)|\Delta| \leqslant \beta \sum \\ \\ S(W_2) & \text{if } (\alpha + \gamma)|\Delta| > \beta \sum \end{cases}$$

where

$$\Delta = -|W_1| + |W_2| \qquad \sum = |W_1| + |W_2|$$

$$\alpha = 0.414 \qquad \gamma = 0.198 \text{ and } \beta = 0.0823$$

5.3.5 Determination of Walsh coefficients for pure sinusoidal waveforms

To apply the previous theorems, one must first determine the Walsh coefficients associated with the fundamental component of the assumed sinusoidal waveform. Consider a voltage waveform, and let its fundamental component $v_1(t)$ be such that

$$v_1(t) = \sqrt{2} F_{1v} \cos \omega_0 t + \sqrt{2} F_{2v} \sin \omega_0 t \tag{5.66}$$

To avoid any confusion that may arise between the Walsh coefficients W_k $(k = 0, 1, \ldots)$ for the whole voltage waveform and those that relate only to the fundamental component of that waveform, we shall denote the Walsh coefficient due to the fundamental by W_{1k} $(k = 0, 1, \ldots)$. It follows from eqn. 5.62 that the fundamental component can be expressed as

$$v_1(t) = \sum_{k=0}^{10} W_{1k, v} \, \text{Wal}\left(k, \frac{t}{T}\right) \tag{5.67}$$

However, the coefficients $W_{11, v}$ and $W_{12, v}$ can be found from their corresponding Fourier coefficients, using the inverse of eqn. 5.50 as

$$W_{11, v} = 0.9 \, F_{1v}$$
$$W_{12, v} = 0.9 \, F_{2v} \tag{5.68}$$

This is so because $F_i = 0$ for all i except $i = 1$ and $i = 2$ (see eqn. 5.46). On the other hand, F_{1v} and F_{2v} can be expressed in terms of the Walsh coefficients due to the whole faulted waveform. Assuming a 10×10 A matrix, F_{1v} and F_{2v} can be found according to eqn. 5.60. By substituting the latter equations into eqn. 5.68, we obtain

$$W_{11, v} = 0.81 \, W_{1, v} - 0.3357 \, W_{5, v} - 0.0666 \, W_{9, v}$$
$$W_{12, v} = 0.81 \, W_{2, v} + 0.3357 \, W_{6, v} - 0.0666 \, W_{10, v} \tag{5.69}$$

It will be noted that $W_{11, v}$ and $W_{12, v}$ are equal to W_1 and W_2 stated in the amplitude and phase theorems.

5.4 References

1 SLEMON, G.R., ROBERTSON, S.D.T., and RAMAMOORTY, M.: 'High speed protection of power systems based on improved power systems models', CIGRE, Paris, June 1986, Paper 31–09
2 RAMAMOORTY, M.: 'Application of digital computers to power system protection', *J. Inst. Eng. India*, 1972, **52**, pp. 235–238
3 McLAREN, P.G., and REDFERN, M.A.: 'Fourier-series techniques applied to distance protection', *Proc. IEE*, 1975, **122**, pp. 1301–1305

4 PHADKE, A.G., HIBKA, T., and IBRAHIM, M.A.: 'A digital computer system for EHV substations: analysis and field test', *IEEE Trans.* 1976, **PAS-95**, pp. 291–301
5 WISZNIEWSKI, A.: 'Signal recognition in protective relaying' in 'Developments in power system protection,' IEE Conf. Publ. 185, 1980, pp. 132–136
6 WISZNIEWSKI, A.: 'How to reduce errors of distance fault locating algorithms', *IEEE Trans.*, 1981, **PAS-100**, pp. 4815–4820
7 JOHNS, A.T., and MARTIN, M.A.: 'Fundamental digital approach to the distance protection of EHV transmission lines', *Proc. IEE* 1978, **125**, pp. 377–384
8 DESOER, C.A., and KUH, E.S.: 'Basic circuit theory', (McGraw-Hill, 1969)
9 HORTON, J.W.: 'The use of Walsh functions for high-speed digital relaying', IEEE PES Summer Meeting, 1975, July 20–25, Paper A 75 582–7

Chapter 6

Least squares based methods

6.1 Introduction

In this Chapter we shall discuss techniques used to fit faulted current and voltage waveforms, each to a sinusoidal waveform containing a fundamental component, a decaying/constant DC component and/or harmonics.

These techniques use the least squares (LSQ) method to minimise the fitting error, and all have the common goal of extracting the fundamental components of voltage and current waveforms, to calculate the impedance to the fault or the comparison of current-based signals in digital differential protection.

6.2 Integral LSQ fit [1]

6.2.1 Basic assumptions

It is assumed that the monitored signal, a voltage waveform $v(t)$, is approximated by a function $g(t)$, which consists of a fundamental component, an exponentially decaying DC component, and/or harmonics. Therefore, in general terms, $g(t)$ can be written as

$$g(t) = K_1 \, e^{-t/\tau} + \sum_{m=1}^{M} (K_{2m} \cos m\omega_0 t + K_{2m+1} \sin m\omega_0 t) \qquad (6.1)$$

where $K_1, K_2, \ldots, K_{2M+1}$ are the unknown constants required to be determined, M is the highest harmonic considered, τ is the time constant of the decaying DC component and ω_0 is the angular frequency of the fundamental component.

6.2.2 Determination of unknown coefficients

The unknown constants of eqn. 6.1 can be determined using the least squares method explained in Section 2.5.1. Consider a voltage waveform $v(t)$. Now the integral of the squared errors S is, as previously explained, given by eqn. 6.2, in which $g(t)$ is the approximation defined in eqn. 6.1:

$$S = \int [v(t) - g(t)]^2 dt \qquad (6.2)$$

When eqns. 6.1 and 6.2 are combined together we get

$$S = \int \left[v(t) - K_1 \, e^{-t/\tau} - \sum_{m=1}^{M} K_{2m}(\cos m\omega_0 t + \sin m\omega_0 t) \right]^2 dt \qquad (6.3)$$

As discussed in Chapter 2, the best fit occurs when S is a minimum, and this is satisfied if

$$\frac{\partial S}{\partial K_i} = 0 \qquad (i = 1, 2, \ldots, 2M+1) \qquad (6.4)$$

By performing the partial derivative with respect to K_1, K_{2m} and K_{2m+1}, we obtain

$$
\left.\begin{aligned}
\frac{\partial S}{\partial K_1} &= 0 = -2 \int [v(t) - K_1\, e^{-t/\tau} \\
&\quad - \sum_{m=1}^{M} (K_{2m} \cos m\omega_0 t + K_{2m+1} \sin m\omega_0 t)]\, e^{-t/\tau} dt \\[2mm]
\frac{\partial S}{\partial K_{2m}} &= 0 = -2 \int [v(t) - K_1\, e^{-t/\tau} \\
&\quad - \sum_{m=1}^{M} (K_{2m} \cos m\omega_0 t + K_{2m+1} \sin m\omega_0 t)]\, \cos m\omega_0 t\; dt \\[2mm]
\frac{\partial S}{\partial K_{2m+1}} &= 0 = -2 \int [v(t) - K_1\, e^{-t/\tau} \\
&\quad - \sum_{m=1}^{M} (K_{2m} \cos m\omega_0 t + K_{2m+1} \sin m\omega_0 t)]\, \sin m\omega_0 t\; dt
\end{aligned}\right\} \qquad (6.5)
$$

The limits of integration in the previous equations are taken from t_1 to $t_1 + T$ (i.e. over one fundamental period). After simplification, eqn. 6.5 then reduces to

$$
\left.\begin{aligned}
K_1 \int_{t_1}^{t_1+T} e^{-2t/\tau} dt &= \int_{t_1}^{t_1+T} v(t)\, e^{-t/\tau} dt \\[2mm]
K_{2m} &= \frac{2}{T} \int_{t_1}^{t_1+T} v(t)\, \cos m\omega_0 t\; dt \\[2mm]
K_{2m+1} &= \frac{2}{T} \int_{t_1}^{t_1+T} v(t)\, \sin m\omega_0 t\; dt
\end{aligned}\right\} \qquad (6.6)
$$

These equations have been derived from eqn. 6.5 using the following identities:

$$\int_{x_1}^{x_1+2} e^{ax} \sin mx\; dx = \frac{1}{a^2 + m^2} [a \sin mx - m \cos mx] e^{ax}]_{x_1}^{x_1+2} = 0$$

$$\int_{x_1}^{x_1+2} e^{ax} \cos mx\; dx = \frac{1}{a^2 + m^2} [a \cos mx - m \sin mx] e^{ax}]_{x_1}^{x_1+2} = 0$$

It must be remembered that the monitored waveform $v(t)$ is available in the form of discrete samples.

Derivation of coefficients in terms of measured waveform samples

If the decaying DC time constant τ is known, then it is possible to express K_1 in terms of the measured samples $S_n = v(t_n)$, $(n = 1, 2, \ldots, N$, where N is the number of samples per cycle). Let $\Delta t =$ sampling interval; then $t_n = (n-1)\Delta t$, which gives $t_1 = 0$, and $T = N\Delta t$ is the fundamental period time.

By approximating the integral using the trapezoidal method, therefore, eqn. 6.6 can be written as

$$\frac{1}{2} K_1 I \Delta t = \frac{1}{2}[v(t_1) \; e^{-t_1/\tau} + 2v(t_2) \; e^{-t_2/\tau} + \cdots + 2v(t_{N-1}) \; e^{-t_{N-1}/\tau} + v(t_N) \; e^{-t_N/\tau}]\Delta t$$

or

$$K_1 = W_{11}S_1 + W_{1N}S_N + \sum_{n=2}^{N-1} 2W_{1n}S_n \qquad (6.7)$$

where

$$I = e^{-2t_1/\tau} + 2 \; e^{-2t_2/\tau} + \cdots + 2 \; e^{-2t_{N-1}/\tau} + e^{-2t_N/\tau}$$

and W_{1n} is the weighting factor of the ith sample used to calculate K_1, which is equal to $e^{-t_i/\tau}/I$.

If the trapezoidal method is applied to the second eqn. 6.6, we obtain

$$K_{2m} = \frac{2}{N\Delta t} \left[\frac{1}{2} v(t_1) \; \cos m\left(\frac{2\pi}{T}\right) t_1 + v(t_2) \; \cos m\left(\frac{2\pi}{T}\right) t_2 + \cdots \right.$$

$$\left. + v(t_{N-1}) \; \cos m\left(\frac{2\pi}{T}\right) t_{N-1} + \frac{1}{2} v(t_N) \; \cos m\left(\frac{2\pi}{T}\right) t_N \right] \Delta t$$

or

$$K_{2m} = \frac{1}{N} \left[W_{2m,\,1}S_1 + W_{2m,\,N}S_N + \sum_{n=2}^{N-1} 2W_{2m,\,n}S_n \right] \qquad (6.8)$$

where

$$W_{2m,\,n} = \text{weighting factor for } n\text{th sample for calculating } K_{2m}$$

$$= \cos m\left(\frac{2\pi}{T}\right) t_n = \cos m\left(\frac{2\pi(n-1)}{N}\right)$$

Similarly K_{2m+1} can be expressed as

$$K_{2m+1} = \frac{1}{N} \left[W_{2m+1,\,1}S_1 + W_{2m+1,\,N}S_N + \sum_{n=2}^{N-1} 2W_{2m+1,\,n}S_n \right] \qquad (6.9)$$

where

$$W_{2m+1,\,n} = \text{weighting factor of } n\text{th sample used to calculate } K_{2m+1}$$

$$= \sin m\left(\frac{2\pi}{T}\right) t_n = \sin m\left(\frac{2\pi(n-1)}{N}\right)$$

6.2.3 Implementation of the algorithm

The above algorithm can be implemented either to extract the fundamental frequency component (as in the case of transmission line protection), or to extract the second-harmonic component as well as the fundamental component. The latter is particularly important when dealing with protection of transformers, as will be discussed in detail in Chapter 10.

6.3 Power series LSQ fit [2]

6.3.1 Basic assumptions

This algorithm assumes that the current and voltage waveforms contain only constant DC and fundamental components. Taking the voltage signal $v(t)$ as an example, then $v(t)$ is assumed to take the form

$$v(t) = K_1 + K_2 \sin(\omega_0 t + \beta) \tag{6.10}$$

6.3.2 Shifted waveform

To minimise the computational burden, the fit is performed using a continuously shifted time reference frame. This may be done by noting that if t^* is the shifted time reference frame and t_s is the time shift, then the new time reference frame t^* is defined as

$$t^* = t - t_s \tag{6.11}$$

If the shifted time reference frame is set to correspond to half the number of sampling intervals $(N-1)\Delta t/2$, then

$$t_s = t - \frac{N-1}{2}\Delta t \tag{6.12}$$

Now consider the samples $S_i (i = 1, 2, \ldots, N)$ occurring at times

$$t - (N-1)\Delta t, \ t - (N-2)\Delta t, \ \ldots, \ t - \frac{(N-1)}{2}\Delta t, \ \ldots \tag{6.13}$$

When these times are transformed to the new reference frame t^*, using eqns. 6.11 and 6.12, the sequence 6.13 becomes

$$-\left(\frac{N-1}{2}\right)\Delta t, \ -\left(\frac{N-1}{2}-1\right)\Delta t, \ \ldots, \ -\Delta t, \ 0, \ \ldots,$$

$$\left(\frac{N-1}{2}-1\right)\Delta t, \ \left(\frac{N-1}{2}\right)\Delta t \tag{6.14}$$

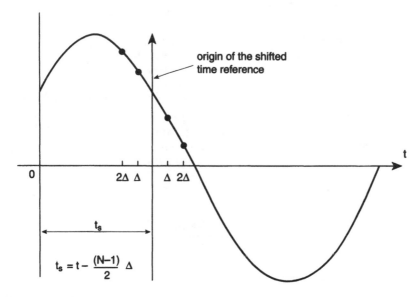

Figure 6.1   Sampled data plotted against a shifted time reference

This sequence shows that the sampling time in the new reference t^*, is independent of time. Figure 6.1 explains the meaning of the new time reference in association with an assumed sinusoidal waveform.

By combining eqns. 6.10 and 6.11, we then obtain

$$v(t) = K_1 + K_2 \sin[\omega_0(t^* + t_s) + \beta]$$

or

$$v(t) = K_1 + K_2 \sin(\omega_0 t^* + \beta^*) \tag{6.15}$$

where

$$\beta^* = \omega_0 t_s + \beta$$

6.3.3   Approximating the shifted waveform by a power series

When eqn. 6.16 is expanded, we obtain

$$v^* = v(t^*) = K_1 + K_2 \sin \omega_0 t^* \cos \beta^* + K_2 \cos \omega_0 t^* \sin \beta^*$$

But

$$\sin \omega_0 t^* \approx \omega_0 t^* - \frac{(\omega_0 t^*)^3}{3!}$$

and

$$\cos \omega_0 t^* \approx 1 - \frac{(\omega_0 t^*)^2}{2!}$$

Therefore the first equation can be expressed as

$$v^* = a_0 + a_1 x + a_2 x^2 + a_3 x^3 \tag{6.16}$$

where

$$x = \omega_0 t^*$$
$$a_0 = K_1 + K_2 \sin \beta^* \qquad\qquad a_1 = K_2 \cos \beta^*$$
$$a_2 = (-K_2 \sin \beta^*)/2! \qquad a_3 = (-K_2 \cos \beta^*)/3!$$

Eqn. 6.16 expresses the sampled value in terms of a-coefficients. In other words, the monitored waveform, which is assumed to consist of a constant term and a fundamental component, is fitted by a power series of third order. The amplitude K_2 and phase angle β of the fundamental component can be expressed in terms of coefficients a_1 and a_2 as follows:

$$K_2 = \sqrt{a_1^2 + 2a_2^2}$$

(6.17)

$$\beta = \tan^{-1}\left(\frac{-2a_2}{a_1}\right) - \omega_0 t_s$$

Determination of the power series coefficients

The a-coefficients of eqn. 6.16 can be determined using the least squares method of Section 2.5.1. If, for example, the measured voltage waveform samples are assumed to be $v_i (i = 1, 2, \ldots, N)$, the sum of the squares of error S can be found using eqns. 2.37 and 2.38 if y_i is replaced by v_i and u by v^*, such that

$$S = \sum_{i=1}^{N} (v_i - v_i^*)^2$$

(6.18)

$$= \sum_{i=1}^{N} [v_i - (a_0 + a_1 x_i + a_2 x_i^2 + a_3 x_i^3)]^2$$

According to eqn. 2.39, the summation of the square of the error S is a minimum if

$$\frac{\partial S}{\partial a_i} = 0 \qquad (i = 0, 1, 2, 3).$$

When this is applied to eqn. 6.18, we obtain the following set of equations:

$$
\begin{bmatrix}
N & \sum x_i & \sum x_i^2 & \sum x_i^3 \\
\sum x_i & \sum x_i^2 & \sum x_i^3 & \sum x_i^4 \\
\sum x_i^2 & \sum x_i^3 & \sum x_i^4 & \sum x_i^5 \\
\sum x_i^3 & \sum x_i^4 & \sum x_i^5 & \sum x_i^6
\end{bmatrix}
\begin{bmatrix}
a_0 \\
a_1 \\
a_2 \\
a_3
\end{bmatrix}
=
\begin{bmatrix}
\sum v_i \\
\sum v_i x_i \\
\sum v_i x_i^2 \\
\sum v_i x_i^3
\end{bmatrix}
$$

(6.19)

Note that the range of summation in eqn. 6.19 is from 1 to N and v_i, x_i are evaluated at times corresponding to $t_i = t - (N - i)\Delta t$.

Note also that only a_1 and a_2 are required to determine the amplitude and phase angle of the fundamental component as per eqn. 6.17. It is, therefore, sufficient to solve eqn. 6.19 only for a_1 and a_2. Furthermore, the computational burden can be greatly reduced if the data are shifted to the new reference frame t^* (which causes x_i to remain constant) and N is chosen to be an odd number. This results in all sums involving odd powers of x becoming zero. Therefore, by manipulating eqn. 6.19, the constants a_1 and a_2 can then be expressed as follows:

$$
\left.
\begin{aligned}
a_1 &= M_1 \sum v_i x_i + M_2 \sum v_i x_i^3 \\[2mm]
a_2 &= M_3 \sum v_i - M_4 \sum v_i x_i^2
\end{aligned}
\right\}
\qquad (6.20)
$$

where

$$
M_1 = \frac{1}{\sum x_i^2} \left(1 - \frac{\left(\sum x_i^4 \right)^2}{\left(\sum x_i^4 \right)^2 - \left(\sum x_i^6 \right)\left(\sum x_i^2 \right)} \right)
$$

$$
M_2 = \frac{\sum x_i^4}{\left(\sum x_i^4 \right)^2 - \left(\sum x_i^6 \right)\left(\sum x_i^3 \right)}
$$

$$
M_3 = \frac{\sum x_i^2}{\left(\sum x_i^2 \right)^2 - N \sum x_i^4}
$$

$$
M_4 = \frac{N}{\left(\sum x_i^2 \right)^2 - N \left(\sum x_i^4 \right)}
$$

The constants M_i, $i = 1, \ldots, 4$ can be determined beforehand once the number of points N, and the sampling interval, have been selected. It should be noted that the limit of the summation is from $i = 1$ to $i = N$.

6.4 Multi-variable series LSQ technique [3]

6.4.1 Basic assumptions

The development of the multivariable series based algorithm is based on the following assumptions:

(i)   In practice, even harmonics are not present in the faulted voltage and current waveforms

(ii) Fifth and higher order harmonics are blocked from reaching the relay by signal preconditioning equipment.

Therefore, it will be assumed that the voltage waveform $v(t)$ consists of a decaying DC component, the fundamental and the third harmonic component, such that

$$v(t) = K_1\, e^{-t/\tau} + K_2\, \sin(\omega_0 t + \theta_1) + K_3\, \sin(3\omega_0 t + \theta_3) \qquad (6.21)$$

6.4.2 Derivation of the multi-variable series

The exponential term $e^{-t/\tau}$ of eqn. 6.21 can be expanded using Taylor's series, such that

$$e^{-t/\tau} = 1 - \frac{t}{\tau} + \frac{1}{2!}\frac{t^2}{\tau^2} - \frac{1}{3!}\frac{t^3}{\tau^3} + \cdots \qquad (6.22)$$

By considering only the first three terms of this expression, eqn. 6.21 can be expressed as

$$v(t) = k_1 - k_1\,\frac{t}{\tau} + k_1\,\frac{t^2}{2\tau^2} + k_{21}\,\sin(\omega_0 t + \theta_1)$$

$$+ k_{23}\,\sin(3\omega_0 t + \theta_3)$$

or

$$v(t) = k_1 - k_1\,\frac{t}{\tau} + k_1\,\frac{t^2}{2\tau^2} + (k_{21}\,\cos\theta_1)\,\sin\omega_0 t + (k_{21}\,\sin\theta_1)\,\cos\omega_0 t$$

$$+ (k_{23}\,\cos\theta_3)\,\sin 3\omega_0 t + (k_{23}\,\sin\theta_3)\,\cos 3\omega_0 t \qquad (6.23)$$

where $k_1 = K_1$, $k_{21} = K_2$, $k_{23} = K_3$.

Therefore, at $t = t_1$ the instantaneous value of the signal can be expressed as

$$v(t_1) = k_1 - k_1\,\frac{t_1}{\tau} + k_1\,\frac{t_1^2}{2\tau^2} + (k_{21}\,\cos\theta_1)\,\sin\omega_0 t_1 + (k_{21}\,\sin\theta_1)\,\cos\omega_0 t_1$$

$$+ (k_{23}\,\cos\theta_3)\,\sin 3\omega_0 t_1 + (k_{23}\,\sin\theta_3)\,\cos 3\omega_0 t_1$$

This equation may be written in the more convenient form

$$S_1 = v(t_1) = a_{11}x_1 + a_{12}x_2 + a_{13}x_3 + a_{14}x_4 + a_{15}x_5 + a_{16}x_6 + a_{17}x_7 \qquad (6.24)$$

where S_1 is the sample measured at time t_1. The coefficients in eqn. 6.24 are related only to the time at which the samples are taken, and they take the form

$$a_{11} = 1 \qquad a_{12} = \sin(\omega_0 t_1) \qquad a_{13} = \cos(\omega_0 t_1)$$
$$a_{14} = \sin(3\omega_0 t_1) \qquad a_{15} = \cos(3\omega_0 t_1)$$
$$a_{16} = t_1 \qquad a_{17} = t_1^2$$

The x-values are functions of the unknowns, and are given by

$$x_1 = k_1 \qquad x_2 = k_{21} \cos \theta_1 \qquad x_3 = k_{21} \sin \theta_1$$
$$x_4 = k_{23} \cos \theta_3 \qquad x_5 = k_{23} \sin \theta_3$$
$$x_6 = -k_1/\tau \qquad x_7 = k_1/(2\tau)^2$$

The sample S_2 measured at time $t = t_2$ can likewise be expressed as

$$S_2 = v(t_2) = a_{21}x_1 + a_{22}x_2 + a_{23}x_3 + a_{24}x_4 + a_{25}x_5 + a_{26}x_6 + a_{27}x_7 \qquad (6.25)$$

where $a_{21} = 1$, $a_{22} = \sin(\omega_0 t_2)$, $a_{23} = \cos(\omega_0 t_2)$, . . ., etc.

As mentioned previously, the a-coefficients are functions of time. Therefore if t_1 is taken as a time reference and the voltage is sampled at preselected times, then the values of the coefficients of eqns. 6.25 and 6.26 can be specified. The values of the sampled voltages S_1 and S_2 are also known so that, by proceeding in this manner, m equations can be generated from m samples of voltages. These equations can be written in matrix form:

$$\underbrace{A}_{m \times 7} \; \underbrace{X}_{7 \times 1} = \underbrace{S}_{m \times 1} \qquad (6.26)$$

The elements of matrix A depend on the time reference and the rate of sampling and, if the number of samples m is taken to be greater than the number of unknowns (i.e. greater than seven), the matrix A is a rectangular matrix which has m rows and seven columns.

The solution of eqn. 6.26 can be found by using the concept of the pseudo-inverse, discussed in Section 2.5.1. This involves premultiplying both sides of eqn. 6.26 by the left pseudoinverse of A, which results in

$$\underbrace{X}_{7 \times 1} = \underbrace{A'}_{7 \times m} \; \underbrace{S}_{m \times 1} \qquad (6.27)$$

where A' is the pseudoinverse of A and is defined as

$$A' = [A^T A]^{-1}[A]$$

After determining the unknown vector X from eqn. 6.27, the magnitude of the fundamental component of voltage K_2 and its phase θ_1 can be found in terms of x_2 and x_3 (see eqns. 6.23, 6.24).

6.5 Determination of measured impedance estimates

The three algorithms covered in this Chapter share the common goal of determining the impedance to the fault using the fundamental components of current and voltage already determined by each of the individual algorithms. According to the integral LSQ fit algorithm, the impedance Z is found as

$$Z = \frac{K_{2V} + jK_{3V}}{K_{2I} + jK_{3I}} \tag{6.28}$$

Now K_{2V}, K_{2I} are the real parts of voltage and current fundamental components, respectively, and are determined using eqn. 6.8 (with $m = 1$). Likewise, K_{3V} and K_{3I} are the imaginary parts of voltage and current fundamental components, respectively, which are determined by using eqn. 6.9 (with $m = 1$).

The measured impedance can alternatively be determined using the power series LSQ fit by means of eqn. 6.29:

$$Z = \frac{K_{2V} \angle \beta_V}{K_{2I} \angle \beta_I} \tag{6.29}$$

In the case of the power series, K_{2V}, K_{2I} are the amplitude of the voltage and current fundamental components, respectively, and β_V, β_I are the phase of the voltage and current fundamental components, respectively. The latter are determined by eqn. 6.17.

Finally, a measured impedance can be found using the multivariable series as

$$Z = \frac{\text{Re}[V] + j\,\text{Im}[V]}{\text{Re}[I] + j\,\text{Im}[I]} = \frac{x_{2V} + jx_{3V}}{x_{2I} + jx_{3I}} \tag{6.30}$$

The values x_{2V}, x_{3V} are the real and imaginary parts of the fundamental of the voltage waveform and are determined according to eqn. 6.27, together with the corresponding values derived for the current (x_{2I}, x_{3I}).

6.6 References

1 LUCKETT, R.G., MUNDAY, P.J., and MURRAY, B.E.: 'A substation based computer for control and protection' Developments in power system protection, IEE Conf. Publ. 125, London, March 1975, pp. 252–260
2 BROOKS, A.W., Jr: 'Distance relaying using least-squares estimates of voltage, current and impedance', Proc. IEEE PICA Conf, 77CH 1131-2, PWR, May 1977, pp. 394–402
3 SACHDEV, M.S., and BARIBEAU, M.A.: 'A new algorithm for digital impedance relays', *IEEE Trans.*, 1979, **PAS-98**, pp. 2232–2240

Chapter 7

Differential equation based techniques

7.1 Introduction

In this Chapter no special assumption will be made with regard to the content of faulted current and voltage waveforms. The fundamental approach, which is common to all algorithms covered in this Chapter, is based on the fact that all protected equipment can be normally represented by differential equations of either first or second order. The methods are described by reference to transmission-line protection, since it is in this application that they are mainly used. However, the methods can easily be extended to other items of plant. For the purposes of this Chapter we shall assume the line length is such that shunt capacitance can either be neglected or can be lumped into a single 'equivalent' value.

7.2 Representation of transmission lines with capacitance neglected

Generally, three-phase transmission lines can be represented by a set of differential equations expressed in terms of basic electrical parameters, namely resistance, inductance and capacitance. Consider first an element of length Δx along a three-phase line (Figure 7.1). If it is assumed that R_k, L_k are the resistance and inductance per unit length of the kth phase, L_{kl} is the mutual

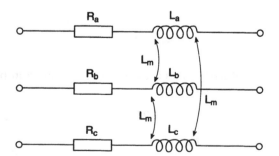

Figure 7.1   Elemental length of distributed transmission line represented by series resistance and self and mutual inductances

inductance per unit length between kth and lth phases and i_k, v_k are the current and voltage of kth phase, then the voltage drop across an infinitesimal length dx of phase 'a' is:

$$dv_a = (R_a dx)i_a + (L_a dx)\frac{di_a}{dt} + (L_{ab}dx)\frac{di_b}{dt} + (L_{ac}dx)\frac{di_c}{dt}$$

$$\frac{dv_a}{dx} = \left(R_a + L_a\frac{d}{dt}\right)i_a + L_{ab}\frac{di_b}{dt} + L_{ac}\frac{di_c}{dt} \tag{7.1}$$

Similarly, the voltage-current relations for phases b and c would be

$$\frac{dv_b}{dx} = L_{ba}\frac{di_a}{dt} + \left(R_a + L_b\frac{d}{dt}\right)i_b + L_{bc}\frac{di_c}{dt} \tag{7.2}$$

$$\frac{dv_c}{dx} = L_{ca}\frac{di_a}{dt} + L_{cb}\frac{di_b}{dt} + \left(R_c + L_c\frac{d}{dt}\right)i_c \tag{7.3}$$

If the line is assumed to be ideally transposed, we have

$$R_a = R_b = R_c = R_s$$

$$L_a = L_b = L_c = L_s \tag{7.4}$$

$$L_{ab} = L_{ac} = L_{ba} = L_{bc} = L_{ca} = L_{cb} = L_m$$

where R_s, L_s are the series resistance and self inductance per unit length of each phase, and L_m is the mutual inductance per unit length between any two phases.

By substituting eqn. 7.4 into eqns. 7.1–7.3, we obtain:

$$\frac{dv_a}{dx} = \left(R_s + L_s\frac{d}{dt}\right)i_a + L_m\frac{di_b}{dt} + L_m\frac{di_c}{dt}$$

$$\frac{dv_b}{dx} = L_m\frac{di_a}{dt} + \left(R_s + L_s\frac{d}{dt}\right)i_b + L_m\frac{di_c}{dt} \tag{7.5}$$

$$\frac{dv_c}{dx} = L_m\frac{di_a}{dt} + L_m\frac{di_b}{dt} + \left(R_s + L_s\frac{d}{dt}\right)i_c$$

Now the self and mutual parameters are related to the zero and positive phase sequence parameters as follows:

$$L_0 = L_s + 2L_m$$

$$R_1 = R_s \tag{7.6}$$

$$L_1 = L_s - L_m$$

and $\qquad\qquad i_0 = (i_a + i_b + i_c)/3$

where R_1, L_1 are positive phase sequence resistance and inductance, respectively, L_0 is the zero sequence inductance and i_0 is the zero sequence current.

By combining eqns. 7.5 and 7.6, we then obtain

$$\frac{dv_a}{dx} = \left(R_1 + L_1 \frac{d}{dt} \right) i_a + (L_0 - L_1) \frac{di_0}{dt}$$

$$\frac{dv_b}{dx} = \left(R_1 + L_1 \frac{d}{dt} \right) i_b + (L_0 - L_1) \frac{di_0}{dt} \tag{7.7}$$

$$\frac{dv_c}{dx} = \left(R_1 + L_1 \frac{d}{dt} \right) i_c + (L_0 - L_1) \frac{di_0}{dt}$$

Eqns. 7.1–7.3, 7.5 and 7.7 can be used to calculate the voltage drop between the fault point and the relay location for different fault types; each of these will be considered in the following subsections.

7.2.1 Single-phase to ground fault

Assume a solid single-phase to ground fault occurs on phase 'a' at a distance x from the relay location. The instantaneous value of the voltage v_a, which is the voltage of phase 'a' at the relaying point, can be calculated using eqns. 7.1 for untransposed lines and eqns. 7.5 or 7.7 for assumed ideally transposed lines.

Using the instantaneous values of the voltages, currents and the rate of change of the currents, the voltage v_a can be obtained using eqn. 7.1 such that

$$v_a = xR_a i_a + xL_a \frac{d}{dt} \left(i_a + \frac{L_{ab}}{L_a} i_b + \frac{L_{ac}}{L_a} i_c \right)$$

or

$$v_a = xR_a i_x + xL_a \frac{di_y}{dt} \tag{7.8}$$

where

$$i_x = i_a$$

and

$$i_y = i_a + (L_{ab}/L_a) i_b + (L_{ac}/L_a) i_c$$

Equations relating to transposed lines can be expressed in the same way, and in this case eqns. 7.5 are used to obtain the relationship given in eqn. 7.9

$$v_a = xR_s i_x + xL_s \frac{di_y}{dt} \tag{7.9}$$

where

$$i_x = i_a$$

and

$$i_y = i_a + (L_m/L_s)i_b + (L_m/L_s)i_c$$

The alternative sequence component formulation of eqns. (7.7) likewise leads to

$$v_a = xR_1 i_x + xL_1 \frac{di_y}{dt} \tag{7.10}$$

where

$$i_x = i_a$$

and

$$i_y = i_a + \frac{L_0 - L_1}{L_1} i_0$$

7.2.2 Phase-to-phase and three-phase faults

When the fault involves two or three phases, the voltage between the faulted phases, say 'a' and 'b', can be found as follows:

$$v_a - v_b = x\left(R_a + L_a \frac{d}{dt}\right)i_a + xL_{ab}\frac{di_b}{dt} - x\left[L_{ba}\frac{di_a}{dt} + \left(R_b + L_b \frac{d}{dt}\right)i_b\right]$$

This equation can be reduced to the more succinct form

$$v_a - v_b = xR_a i_x + x(L_a - L_{ab})\frac{di_y}{dt} \tag{7.11}$$

where

$$i_x = i_a - (R_b/R_a)i_b$$

and

$$i_y = i_a - \frac{(L_b - L_{ab})i_b}{L_a - L_{ab}}$$

When the line is assumed ideally transposed, eqn. 7.10 can be written in terms of the difference between the currents on each faulted phase, giving

$$v_a - v_b = xR_1 i_x + xL_1 \frac{di_y}{dt} \tag{7.12}$$

where in this case $i_x = i_y = i_a - i_b$.

The above analysis shows that the behaviour of the transmission line under fault conditions is governed by a differential equation having the general form of eqn. 7.13. The measured values of the currents and voltages are taken in the form of samples and the measurement is usually done simultaneously on all three phases by using suitable sample-and-hold peripheral equipment as described previously.

$$v = Ri_x + L\frac{di_y}{dt} \tag{7.13}$$

The values of R and L in eqn. 7.13 can be calculated by several methods, described in more detail below.

7.3 Differential equation protection with selected limits [1–3]

7.3.1 Basic principles

If eqn. 7.13 is integrated once over the time interval t_1 to t_2 and again over the period from t_3 to t_4, the following equations are obtained:

$$R \int_{t_1}^{t_2} i_x dt + L(i_{y2} - i_{y1}) = \int_{t_1}^{t_2} v \, dt \tag{7.14}$$

$$R \int_{t_3}^{t_4} i_x dt + L(i_{y4} - i_{y3}) = \int_{t_3}^{t_4} v \, dt \tag{7.15}$$

It will be evident that the values of measured resistance R and inductance L can be determined by solving these two equations. However, due to the presence of harmonics, particularly in the current waveforms during the first cycle after the occurrence of the fault, it has been found that the values of R and L calculated using this simple approach are very sensitive to low-frequency harmonics. The accuracy of calculated values of R and L can, however, be greatly improved if the limits of integration are so chosen as to eliminate (or filter out) the unwanted harmonics.

7.3.2 Digital harmonic filtering by selected limits

Harmonic components within waveforms can be eliminated if the waveform in question is integrated over appropriate periods. Let us consider, for example, a current waveform $i(t)$, which will be assumed periodic in the interval from $t = t_1$ to $t_1 + T$. Using the previously described Fourier expansion of eqn. 2.59, $i(t)$ can be expressed as

$$i(t) = \frac{a_0}{2} + a_1 \cos \omega_0 t + a_2 \cos 2\omega_0 t + a_3 \cos 3\omega_0 t + \cdots$$

$$+ b_1 \sin \omega_0 t + b_2 \sin 2\omega_0 t + b_3 \sin 3\omega_0 t + \cdots \tag{7.16}$$

If the highest harmonic contained in the waveform $i(t)$ is N, the last equation would be reduced to

$$i(t) = c_0 + \sum_{m=1}^{N} c_m \cos(m\omega_0 t + \theta_m) \qquad (7.17)$$

where

$$c_0 = \frac{a_0}{2},$$

$$c_m = \sqrt{a_m^2 + b_m^2}$$

and

$$\theta_m = \tan^{-1} \frac{b_m}{a_m}$$

Now let us integrate eqn. 7.17 from $t_1 = 0$ to $t_2 = \alpha/\omega_0$, and let us call this integration I_1 so that:

$$I_1 = \int_{t_1=0}^{t_2=\alpha/\omega_0} i(t) \; dt = \int_0^{\alpha/\omega_0} c_0 dt + \sum_{m=1}^{N} \int_0^{\alpha/\omega_0} c_m \cos(m\omega_0 t + \theta_m) dt$$

This equation shows that the integration of $i(t)$ with respect to time is equal to the summation of the integration of its individual harmonic components. Therefore, let us concentrate our attention on the integration of the mth harmonic component and its multiples. If I_{n1} is the integration of the nth harmonic over the period from $t_1 = 0$ to $t_2 = \alpha/\omega_0$, then

$$I_{n1} = \int_0^{\alpha/\omega_0} c_n \cos(n\omega_0 t + \theta_n) dt$$

which gives

$$I_{n1} = \frac{c_n}{n\omega_0} [\sin(n\alpha + \theta_n) - \sin \theta_n] \qquad (7.18)$$

Similarly, if we integrate $i(t)$ from $t_3 = \pi/n\omega_0$ to $t_4 = (\pi/n + \alpha)/\omega_0$, and if I_{n2} is the corresponding integration of the nth harmonic, then

$$I_{n2} = \int_{\pi/n\omega_0}^{(\pi/n+\alpha)/\omega_0} c_n \cos(n\omega_0 t + \theta_n) dt$$

$$= \frac{c_n}{n\omega_0} [-\sin(n\alpha + \theta_n) + \sin \theta_n] \qquad (7.19)$$

It will be evident from eqns. 7.18 and 7.19 that the sum $I_{n1} + I_{n2} = 0$. This is of course true for all harmonics (n). In essence, this means that any nth harmonic and its multiples can be filtered out from the waveform $i(t)$, by simply adding two integrals taken once over the limits $t_1 = 0$ to $t_2 = \alpha/\omega_0$ and $t_3 = (\pi/n)/\omega_0$ to $t_4 = (\pi/n + \alpha)/\omega_0$.

7.3.3 Graphical interpretation of digital filtering by integration over selected limits

The filtering of harmonics from waveforms by using integration over selected limits can be usefully explained graphically. Consider a waveform $i(t)$ and assume that it consists of only fundamental and third harmonic components, which can in turn be described by

$$i(t) = i_1(t) + i_3(t)$$
$$= I_1 \cos(\omega_0 t + \theta_1) + I_3 \cos(3\omega_0 t + \theta_3)$$

When $i(t)$ is integrated from $t_1 = 0$ to $t_2 = \alpha/\omega_0$ and from $t_3 = (\pi/3)/\omega_0$ to $t_4 = (\pi/3 + \alpha)/\omega_0$ the corresponding integrations of the third harmonic components are equal to the areas shown shaded in Figure 7.2. These areas are equal to each other but have opposite signs and, when added together, they are cancelled and therefore effectively filtered out from the original waveform $i(t)$.

7.3.4 Filtering of multiple harmonic components

We have seen how it is possible to eliminate, or filter out, nth harmonic components by simply adding two integrations together. This is because the nth harmonic contained in the first integration period is cancelled out if the second integration is performed over limits that are related to nth harmonic order. The same basic approach can be used to eliminate any other harmonic m. This is achieved by simply adding a third integration over limits related to mth order harmonic. Therefore, the limits under this condition would be from $t_5 = (\pi/m)/\omega_0$ to $t_6 = (\pi/m + \alpha)/\omega_0$.

The equation used to eliminate nth and mth harmonics simultaneously from a waveform $i(t)$ would then be as follows:

$$\int_0^{\alpha/\omega_0} i(t)dt + \int_{\pi/n\omega_0}^{(\alpha + \pi/n)/\omega_0} i(t)dt + \int_{\pi/m\omega_0}^{(\alpha + \pi/m)/\omega_0} i(t)dt \tag{7.20}$$

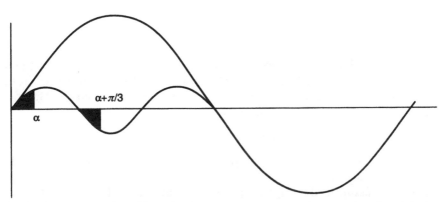

Figure 7.2 Physical interpretation of digital filtering by integration over selected limits

However, the number of integrations can be reduced to only two if the value of α is chosen to be equal to the angle corresponding to a full cycle of the mth harmonic order (i.e. $\alpha = 2\pi/m$). In this way we can ensure the elimination of the effect of the mth harmonic by the first integration, as the integration over a full cycle of a sinusoid is always equal to zero. It is then only required to eliminate the effect of the nth harmonic, which can be done by addition of a second integration. Thus, for removing two harmonics of order n and m together with any multiples thereof, the following equation would be used:

$$\int_0^{2\pi/m\omega_0} i(t)\,dt + \int_{\pi/n\omega_0}^{(\pi/n+2\pi/m)/\omega_0} i(t)\,dt \tag{7.21}$$

By applying the above described principles to eqn. 7.13, it is possible to calculate R and L so that any number of harmonics are eliminated. For example, in order to remove the third and fifth harmonics, eqn. 7.13 can be integrated over

$$\left(0, \frac{2\pi/5}{\omega_0}\right)$$

and again over the interval

$$\left[\frac{\pi/3}{\omega_0}, \left(\frac{\pi/3 + 2\pi/5}{\omega_0}\right)\right].$$

The resulting equation, when added, gives

$$L\left[\int_0^{2\pi/5\omega_0} di_y + \int_{\pi/3\omega_0}^{(\pi/3+2\pi/5)/\omega_0} di_y\right] + R\left[\int_0^{2\pi/5\omega_0} i_x\,dt + \int_{\pi/3\omega_0}^{(\pi/3+2\pi/5)/\omega_0} i_x\,dt\right]$$

$$= \left[\int_0^{2\pi/5\omega_0} v\,dt + \int_{\pi/3\omega_0}^{(\pi/3+2\pi/5)/\omega_0} v\,dt\right] \tag{7.22}$$

In general, this principle can be extended to any number of harmonics by making a sufficient number of integrations.

7.4 Simultaneous differential equation techniques [4, 5]

When these techniques are used, the current and voltage at the relaying point are related by differential equations. The fundamental assumption used in this approach is that the transmission line can be represented by either a lumped-series impedance or a single PI-section.

7.4.1 Lumped series impedance based algorithms

This algorithm assumes that the current and voltage waveforms contain a DC component but are otherwise free from high-frequency oscillations. In other words, it is assumed that high-frequency oscillations are filtered out from the original faulted waveforms using low-pass filtering. This is equivalent to saying

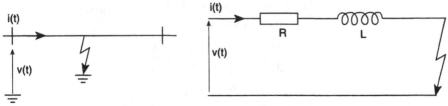

Figure 7.3   Representation of transmission line using lumped-series circuit parameters

that the filtered voltages and currents are produced by a lumped-series transmission-line model, such as that shown in Figure 7.3, where v, i are voltage and current waveforms at the relaying point and R, L are the total resistance and inductance of the line.

Therefore from the basic principles of circuit theory, $v(t)$ and $i(t)$ are related by

$$v = Ri + L\frac{di}{dt} \qquad (7.23)$$

To determine the values of R and L using this equation, it is basically required to have at least two sets of voltage and current samples. Let v_k, i_k be the voltage and current samples at time t_k and let Δt be the sampling time interval. The derivative of the current with respect to time (di/dt) can be approximately determined in sample data form by using eqn. 2.33 to obtain

$$\frac{di}{dt} = \frac{i_{k+1} - i_{k-1}}{2\Delta t} \qquad (7.24)$$

Therefore, using sample notations, eqn. 7.23 becomes

$$v_k \simeq Ri_k + L\frac{(i_{k+1} - i_{k-1})}{2\Delta t} \qquad (7.25)$$

Similarly, by using the following sets of samples at t_{k+1}, eqn. 7.23 becomes

$$v_{k+1} \simeq Ri_{k+1} + L\frac{i_{k+2} - i_k}{2\Delta t} \qquad (7.26)$$

In matrix form, eqns. 7.25 and 7.26 can be combined to give

$$\begin{bmatrix} i_k & \dfrac{i_{k+1} - i_{k-1}}{2\Delta t} \\ i_{k+1} & \dfrac{i_{k+2} - i_k}{2\Delta t} \end{bmatrix} \begin{bmatrix} R \\ L \end{bmatrix} = \begin{bmatrix} v_k \\ v_{k+1} \end{bmatrix} \qquad (7.27)$$

It is evident that this equation can be written in the following short form:

$$AP = V \qquad (7.28)$$

where

$$A = \begin{bmatrix} i_k & \dfrac{i_{k+1} - i_{k-1}}{2\Delta t} \\ i_{k+1} & \dfrac{i_{k+2} - i_k}{2\Delta t} \end{bmatrix}, \quad P = \begin{bmatrix} R \\ L \end{bmatrix} \text{ and } V = \begin{bmatrix} v_k \\ v_{k+1} \end{bmatrix}$$

By solving eqn. 7.27 or 7.28, the parameters R, L can be expressed in terms of the current and voltage samples as follows:

$$R \approx \frac{v_k(i_{k+2} - i_k) - v_{k+1}(i_{k+1} - i_{k-1})}{i_k(i_{k+2} - i_k) - i_{k+1}(i_{k+1} - i_{k-1})} \tag{7.29}$$

$$L \approx 2\Delta t \, \frac{(i_k v_{k+1} - i_{k+1} v_k)}{i_k(i_{k+2} - i_k) - i_{k+1}(i_{k+1} - i_{k-1})} \tag{7.30}$$

It must be remembered that this algorithm produces a succession of estimates of R and L as the sampled data is acquired, and these estimates can be compared for convergence on any post fault values.

7.4.2 Single PI section transmission line model based algorithms

These algorithms take a more fundamental approach with regard to the inclusion of high-frequency oscillations, which can occur during faults. This can be achieved, at least to a limited degree, by including the effect of capacitance of the transmission line using a single PI section transmission line model (Figure 7.4).

7.4.3 Development of the algorithm and basic assumptions

Throughout the derivation of this algorithm, it will be assumed that the magnitude of any arc resistance R_f is so small that the effect of the far end capacitance C can be neglected. Therefore, with reference to Figure 7.4, the circuit equation can be formulated as follows:

$$R(i - i_c) + L \frac{d(i - i_c)}{dt} = v \tag{7.31}$$

The current i_c is the capacitive current flowing through the lumped equivalent capacitance at the relay location, and it is related simply to the voltage at the relaying point by noting that $i_c = C dv/dt$. Eqn. 7.31 thus takes a second-order form:

$$Ri + L \frac{di}{dt} - RC \frac{dv}{dt} - LC \frac{d^2v}{dt^2} = v \tag{7.32}$$

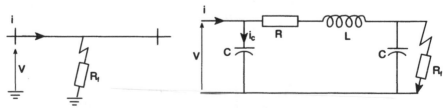

Figure 7.4   Single PI circuit transmission-line model

To determine R and L from eqn. 7.32 it is necessary to use at least four sets of voltage and current samples. Therefore, let us assume that at $t = t_k$ we have $v = v_k$ and $i = i_k$. Also, by using central finite differences, the current and voltage derivatives can be determined from the measured samples using eqns. 2.33 and 2.34 as follows:

$$\left. \begin{aligned} \left.\frac{di}{dt}\right|_{t=t_k} &= \frac{i_{k+1} - i_{k-1}}{2\Delta t} \\[2mm] \left.\frac{dv}{dt}\right|_{t=t_k} &= \frac{v_{k+1} - v_{k-1}}{2\Delta t} \end{aligned} \right\} \tag{7.33}$$

$$\left.\frac{d^2v}{dt^2}\right|_{t=t_k} = \frac{v_{k+1} - 2v_k + v_{k-1}}{(\Delta t)^2} \tag{7.34}$$

Substituting current and voltage samples at $t = t_k$ and their corresponding derivatives described by eqns. 7.33 and 7.34, we obtain

$$Ri_k + L\frac{(i_{k+1} - i_{k-1})}{2\Delta t} - RC\frac{(v_{k+1} - v_{k-1})}{2\Delta t} - LC\frac{(v_{k+1} - 2v_k + v_{k-1})}{(\Delta t)^2} = v_k$$

Similarly, at instants of time t_{k+1}, t_{k+2} and t_{k+3}, we obtain

$$Ri_{k+1} + L\frac{(i_{k+2} - i_k)}{2\Delta t} - RC\frac{(v_{k+2} - v_k)}{2\Delta t} - LC\frac{(v_{k+2} - 2v_{k+1} + v_k)}{(\Delta t)^2} = v_{k+1}$$

$$Ri_{k+2} + L\frac{(i_{k+3} - i_{k+1})}{2\Delta t} - RC\frac{(v_{k+3} - v_{k+1})}{2\Delta t} - LC\frac{(v_{k+3} - 2v_{k+2} + v_{k+1})}{(\Delta t)^2} = v_{k+2}$$

$$Ri_{k+3} + L\frac{(i_{k+4} - i_{k+2})}{2\Delta t} - RC\frac{(v_{k+4} - v_{k+2})}{2\Delta t} - LC\frac{(v_{k+4} - 2v_{k+3} + v_{k+2})}{(\Delta t)^2} = v_{k+3}$$

The above equations can be written in the matrix form

$$\begin{bmatrix} i_k & \dfrac{i_{k+1} - i_{k-1}}{2\Delta t} & -\dfrac{v_{k+1} - v_{k-1}}{2\Delta t} & -\dfrac{v_{k+1} + 2v_k + v_{k-1}}{(\Delta t)^2} \\[4mm] i_{k+1} & \dfrac{i_{k+2} - i_k}{2\Delta t} & -\dfrac{v_{k+2} - v_k}{2\Delta t} & -\dfrac{v_{k+2} - 2v_{k+1} + v_k}{(\Delta t)^2} \\[4mm] i_{k+2} & \dfrac{i_{k+3} - i_{k+1}}{2\Delta t} & -\dfrac{v_{k+3} - v_{k+1}}{2\Delta t} & -\dfrac{v_{k+3} - 2v_{k+2} + v_{k+1}}{(\Delta t)^2} \\[4mm] i_{k+3} & \dfrac{i_{k+4} - i_{k+2}}{2\Delta t} & -\dfrac{v_{k+4} - v_{k+2}}{2\Delta t} & -\dfrac{v_{k+4} - 2v_{k+3} + v_{k+2}}{(\Delta t)^2} \end{bmatrix} \begin{bmatrix} R \\[4mm] L \\[4mm] CR \\[4mm] CL \end{bmatrix} = \begin{bmatrix} v_k \\[4mm] v_{k+1} \\[4mm] v_{k+2} \\[4mm] v_{k+3} \end{bmatrix} \tag{7.35}$$

In short form, eqn. 7.35 becomes:

$$\begin{bmatrix} A & B \\ C & D \end{bmatrix} \begin{bmatrix} X \\ CX \end{bmatrix} = \begin{bmatrix} V1 \\ V2 \end{bmatrix} \tag{7.36}$$

where A, B, C and D are 2×2 submatrices of the 4×4 coefficient matrix in eqn. 7.35

$$X = \begin{bmatrix} R \\ L \end{bmatrix}, \ CX = \begin{bmatrix} CR \\ CL \end{bmatrix}, \ V1 = \begin{bmatrix} v_k \\ v_{k+1} \end{bmatrix} \text{ and } V2 = \begin{bmatrix} v_{k+2} \\ v_{k+3} \end{bmatrix}$$

Since the vectors X and CX are linearly related, eqn. 7.36 can be reduced to

$$(A - BD^{-1}C)X = (V1 - BD^{-1}V2) \tag{7.37}$$

which can be written in a short form

$$FX = G \tag{7.38}$$

The vector X in eqn. 7.38 is readily evaluated by forming the matrices F and G from sampled values. Again this process can be repeated after each sample set becomes available, thus producing a succession of estimates of resistance and inductance between the relaying and fault points.

7.5 References

1 RANJBAR, A.M., and CORY, B.J.: 'Algorithm for distance protection' Developments in power system protection, IEE Conf. Publ. 125, London, March 1975, pp. 276–283
2 RANJBAR, A.M., and CORY, B.J. 'An improved method for the digital protection of high voltage transmission lines', *IEEE Trans.* 1975, **PAS-94**, pp. 544–550
3 GILBERT, J.G., UDREN, E.A., and SACKIN, M.: 'Evaluation of algorithms for computer relaying', IEEE PES Summer Meeting, 1977 Mexico City, Paper A 77 520–0
4 SMOLINSKI, W.J.: 'An algorithm for digital impedance calculation using a single PI section transmission line model, *IEEE Trans.*, 1979, **PAS-98**, pp. 1546–1551
5 JEYASURAY, B., and SMOLINSKI, W.J.: 'Identification of a best algorithm for digital distance protection of transmission lines', *ibid.*, 1983, **PAS-102**, pp. 3358–3359

Chapter 8

Fundamentals of travelling-wave based protection

8.1 Introduction

In previous chapters, digital algorithms were derived on the assumption that the current and voltage waveforms are either sinusoids of a single frequency or sinusoids consisting of decaying/constant DC, fundamental and/or harmonic components. Many of these assumptions are valid and lead to an acceptable performance in applications where the line length is limited. However, the introduction of long EHV/UHV transmission lines and EHV cables into power systems can produce difficult problems for protective relays. The principal causes of these problems are:

(i)   shunt leakage current due to the increase in shunt capacitance on long lines and cables

(ii) relatively low frequency transient currents caused by surge travelling waves following fault inception

(iii) transient currents, at frequencies as low as a few hundred Hertz, caused by the interaction between the total inductance and capacitance of the system

(iv) a weakly damped transient DC current caused by high system reactance-to-resistance ratio.

The foregoing considerations, together with a desire to reduce fault-clearance times to improve the transient stability of electric power systems, has led to much interest in the development of so-called travelling-wave protection.

In this Chapter, transmission lines are treated as distributed circuits, in order to explain how travelling-wave phenomena are propagated and detected. The basic principles of travelling-wave schemes, the formation of relaying signals, Bergeron's equation and discriminant functions using single-phase line models are also explained. These principles are then extended to three-phase lines by decomposing the line into three equivalent single-phase lines using the modal decomposition approach.

8.2 The transmission line as a distributed component [1, 2]

Travelling-wave methods are usually more suitable for application to long lines (typically lines with lengths of more than 250 km). Under such circumstances

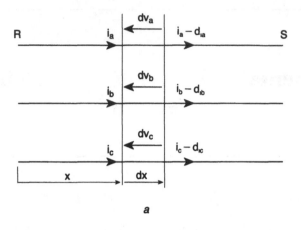

a

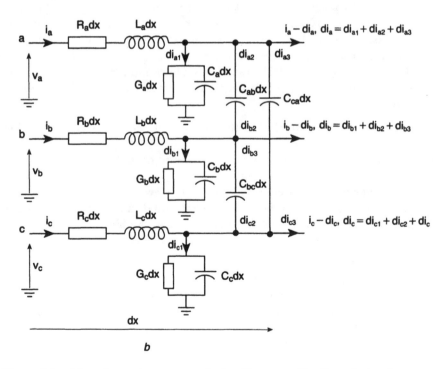

b

Figure 8.1   (*a*) a three-phase transmission line as a distributed circuit
(*b*) representation of an element *dx*

the effect of capacitance has to be included in the representation of the line as a distributed element.

For simplicity, let us ignore losses in the earth return and consider an elemental length of the three-phase line shown in Figure 8.1. Assume that R_k, G_k, L_k, C_k are the self-resistance, conductance, inductance and capacitance per unit length of phase k (where $k=a$, b and c) and L_{km}, C_{km} are the mutual inductance and capacitance, respectively, per unit length between kth and mth phases.

If the line is subjected to transient conditions such as faults, the associated changes in the voltages and currents within an element dx of the line (which is a distance x from the relay location) at an instant t are related by the following equations:

For the 'a' phase:

$$dv_a = \frac{\partial v_a}{\partial x}\, dx = \left(R_a dx + L_a dx\, \frac{\partial}{\partial t} \right) i_a + L_{ab} dx\, \frac{\partial i_b}{\partial t} + L_{ca} dx\, \frac{\partial i_c}{\partial t}$$

$$di_a = \frac{\partial i_a}{\partial x}\, dx = \left(G_a dx + C_a dx\, \frac{\partial}{\partial t} \right) v_a + C_{ab} dx\, \frac{\partial v_{ab}}{\partial t} + C_{ca} dx\, \frac{\partial v_{ac}}{\partial t}$$

By considering the line transient at an instant of time t (i.e. $dt=0$), we can derive the current and voltage relationships for phase 'a':

$$\frac{\partial v_a}{\partial x} = \left(R_a + L_a\, \frac{\partial}{\partial t} \right) i_a + L_{ab}\, \frac{\partial i_b}{\partial t} + L_{ca}\, \frac{\partial i_c}{\partial t}$$

$$\frac{\partial i_a}{\partial x} = \left(G_a + C_a\, \frac{\partial}{\partial t} \right) v_a + C_{ab}\, \frac{\partial v_{ab}}{\partial t} + C_{ca}\, \frac{\partial v_{ac}}{\partial t} \tag{8.1}$$

Similar equations can be derived for phases 'b' and 'c':
For phase 'b':

$$\left. \begin{aligned} \frac{\partial v_b}{\partial x} &= L_{ab}\, \frac{\partial i_a}{\partial t} + \left(R_b + L_{ab}\, \frac{\partial}{\partial t} \right) i_b + L_{bc}\, \frac{\partial i_c}{\partial t} \\[2mm] \frac{\partial i_b}{\partial x} &= C_{ab}\, \frac{\partial v_{ba}}{\partial t} + \left(G_b + C_b\, \frac{\partial}{\partial t} \right) v_b + C_{bc}\, \frac{\partial v_{bc}}{\partial t} \end{aligned} \right\} \tag{8.2}$$

For phase 'c':

$$\left. \begin{aligned} \frac{\partial v_c}{\partial x} &= L_{ca}\, \frac{\partial i_a}{\partial t} + L_{bc}\, \frac{\partial i_b}{\partial t} + \left(R_c + L_c\, \frac{\partial}{\partial t} \right) i_c \\[2mm] \frac{\partial i_c}{\partial x} &= C_{ca}\, \frac{\partial v_{ca}}{\partial t} + C_{bc}\, \frac{\partial v_{cb}}{\partial t} + \left(G_c + C_c\, \frac{\partial}{\partial t} \right) v_c \end{aligned} \right\} \tag{8.3}$$

By replacing v_{ab}, v_{ac}, v_{ba}, v_{bc}, v_{ca}, v_{cb}, by their equivalent phase voltages (i.e. $v_{ab} = v_a - v_b$, . . . , etc.), eqns. 8.1–8.3 can be written:

$$\frac{\partial v}{\partial x} = zi$$

$$\frac{\partial i}{\partial x} = yv \qquad (8.4)$$

where

$$v = \begin{bmatrix} v_a \\ v_b \\ v_c \end{bmatrix} \qquad i = \begin{bmatrix} i_a \\ i_b \\ i_c \end{bmatrix}$$

and

$$z = \begin{bmatrix} R_a + L_a \dfrac{\partial}{\partial t} & L_{ab} \dfrac{\partial}{\partial t} & L_{ca} \dfrac{\partial}{\partial t} \\[2ex] L_{ab} \dfrac{\partial}{\partial t} & R_b + L_b \dfrac{\partial}{\partial t} & L_{bc} \dfrac{\partial}{\partial t} \\[2ex] L_{ca} \dfrac{\partial}{\partial t} & L_{bc} \dfrac{\partial}{\partial t} & R_c + L_c \dfrac{\partial}{\partial t} \end{bmatrix}$$

$$y = \begin{bmatrix} G_a + C_{aa} \dfrac{\partial}{\partial t} & C_{ab} \dfrac{\partial}{\partial t} & C_{ca} \dfrac{\partial}{\partial t} \\[2ex] C_{ab} \dfrac{\partial}{\partial t} & G_b + C_{bb} \dfrac{\partial}{\partial t} & C_{bc} \dfrac{\partial}{\partial t} \\[2ex] C_{ca} \dfrac{\partial}{\partial t} & C_{bc} \dfrac{\partial}{\partial t} & G_c + C_{cc} \dfrac{\partial}{\partial t} \end{bmatrix}$$

where

$$C_{aa} = C_a + C_{ab} + C_{ca}$$

$$C_{bb} = C_{ab} + C_b + C_{bc}$$

$$C_{cc} = C_{ca} + C_{bc} + C_c$$

By replacing the time operator $\partial/\partial t$ in eqns. 8.4 with the transform operator, the latter can be written as

$$\frac{\partial v}{\partial x} = Zi$$

$$\frac{\partial i}{\partial x} = Yv \qquad (8.5)$$

where $\mathbf{Z}$ and $\mathbf{Y}$ are the same as z and y except that the time operator $\partial/\partial t$ has been replaced by the transform operator p. By combining eqns. 8.5 we finally obtain

$$\frac{\partial^2 v}{\partial x^2} = \mathbf{P} v \tag{8.6}$$

$$\frac{\partial^2 i}{\partial x^2} = \mathbf{P}^T i \tag{8.7}$$

where

$$\mathbf{P} = \mathbf{ZY}$$

$$\mathbf{P}^T = \mathbf{YZ}$$

For the purpose of this and the following Chapters, the formulation of eqns. 8.6 and 8.7 is particularly useful for an appreciation of the important special cases of single-phase lines and three-phase transposed lines.

8.2.1 Travelling waves in assumed lossless single-phase lines

Power lines are of course normally of the three-phase type. However, it is much simpler to understand travelling-wave concepts and associated methods by first considering wave propagation in single-phase lines. The equations describing single-phase lines can be obtained directly from eqn. 8.5 by taking their scalar version:

$$\frac{\partial v}{\partial x} = Zv$$

$$\frac{\partial i}{\partial x} = Yi \tag{8.8}$$

where v and i are the voltage and line current, respectively, measured at any point on the line and $Z = R + Lp$, $Y = G + Cp$. In this case the previously defined line parameters R, G, L, C represent single-phase line parameters per unit length.

When eqns. 8.8 are combined, they produce the scalar form of eqns. 8.6 and 8.7, such that

$$\frac{\partial^2 v}{\partial x^2} = \gamma^2 v$$

$$\frac{\partial^2 i}{\partial x^2} = \gamma^2 i \tag{8.9}$$

where γ is known as the propagation constant, and its value is given by

$$\gamma^2 = ZY = (R + Lp)(G + Cp)$$

By solving eqn. 8.9 as an ordinary differential equation in x and then finding i, we obtain

$$v(x) = k_1 \, e^{-\gamma x} + k_2 \, e^{\gamma x}$$

and

$$i(x) = \frac{1}{\sqrt{X/Y}} \, (k_1 \, e^{-\gamma x} - k_2 \, e^{\gamma x})$$

where k_1 and k_2 are constants.

These equations are valid only at some fixed time, i.e. it is assumed that $v(x)$ and $i(x)$ are functions of the distance x along the line. However, since the line voltage and current are functions of the time t as well as the distance x, the equations need to be modified by replacing $v(x)$ and $i(x)$ with $v(x, t)$ and $i(x, t)$, and the constants k_1 and k_2 by the time functions $F_1(t)$ and $F_2(t)$, such that

$$v(x, t) = e^{-\gamma x} F_1(t) + e^{+\gamma x} F_2(t) \tag{8.10}$$

$$i(x, t) = \frac{1}{\sqrt{Z/Y}} \, (e^{-\gamma x} F_1(t) - e^{\gamma x} F_2(t)) \tag{8.11}$$

For an assumed lossless line, $R = G = 0$, which in turn leads to

$$\sqrt{ZY} = p\sqrt{LC} = \frac{p}{c} = \gamma \text{ and}$$

$$\sqrt{\frac{Z}{Y}} = \sqrt{\frac{L}{C}} = Z_0$$

$$v(x, t) = e^{-xp/c} F_1(t) + e^{+xp/c} F_2(t) \tag{8.12}$$

$$i(x, t) = (e^{-xp/c} F_1(t) - e^{+xp/c} F_2(t))/Z_0 \tag{8.13}$$

where

$$c = \frac{1}{\sqrt{LC}}$$

is known as the velocity of propagation and

$$Z_0 = \sqrt{\frac{L}{C}}$$

is known as the surge impedance.

It is possible, with the help of Taylor's theorem, to show that

$$e^{\pm ap} f(t) = f(t \pm a)$$

If this relation is applied to eqns. 8.12 and 8.13 we obtain

$$v(x, t) = F_1\left(t - \frac{x}{c}\right) + F_2\left(t + \frac{x}{c}\right) \tag{8.14}$$

$$i(x, t) = \left(F_1\left(t - \frac{x}{c}\right) - F_2\left(t + \frac{x}{c}\right)\right)\Big/ Z_0 \tag{8.15}$$

Functions $F_1(t - x/c)$ and $F_2(t + x/c)$ thus represent travelling waves in the forward and backward directions of x, respectively, and eqns. 8.14 and 8.15 may therefore be written in the simple terms

$$v = v^+ + v^-$$
$$i = i^+ + i^-$$

(8.16)

where $v^+ = F_1(t - x/c)$, and $v^- = F_2(t + x/c)$ are the forward and backward voltage component, respectively. The values of i^+ and i^- are similarly the forward and backward components of current.

The physical meaning of the function having the form $f(t \pm x/c)$, can be well understood if consideration is given to the behaviour of this function when the argument is held constant, i.e. let $t - x/c = k_1$ for a forward travelling wave and $t + x/c = k_2$ for a backward travelling wave. Therefore any change in time t, say Δt, requires a change in x equal to Δx such that $\Delta x = c\Delta t$ for the forward waveform and $\Delta x = -c\Delta t$ for the backward waveform. Such considerations show that a forward waveform travels in the positive direction of x, while the backward waveform travels in the negative direction of x.

The forward and backward components are related to each other, as seen from eqns. 8.14 through 8.16, by the characteristic (or surge) impedance of the line as follows:

$$v^+ = Z_0 i^+$$
$$v^- = -Z_0 i^-$$

(8.17)

It will be seen from the above analysis that the surge impedance (Z_0) is a real number for an assumed lossless line and it is evident from eqns. 8.17 that the current components are simply a replica of their corresponding voltages. They also show that, while the forward voltage and current waveforms are of the same sign, the backward voltage and current waveforms are of opposite sign as illustrated in Figure 8.2.

Coefficient of reflection
Waves travelling over assumed homogeneous lossless lengths of transmission line continue to propagate at a uniform velocity c and are unchanged in shape. However, at points of discontinuity, such as open circuits or other line terminations, part of the incident wave is reflected back along the line and part is transmitted into and beyond the discontinuity. The wave impinging on the discontinuity is often called an incident wave and the two waves to which it gives rise are normally referred to as reflected and transmitted waves.

Consider Figure 8.3, which shows a transmission line whose sending end is connected to a source and its receiving end is connected to an impedance Z_r. When the switch S is closed, the voltage wave v travels along the line until it reaches the discontinuous receiving-end terminal where only part of the wave passes to the terminal impedance Z_r (transmitted wave), and the rest is reflected back. Let us call the first part v^+ and the second part v^-. Therefore, at the receiving end the total voltage and current are equal to the sum of the forward and backward components of voltage and current:

$$v = v^+ + v^-$$
$$i = i^+ + i^-$$

(8.18)

We recall from the previous Section that the voltage wave and its associated current are interrelated by the surge (or characteristic) impedance of the line:

$$i^+ = \frac{v^+}{Z_0}$$

(8.19)

$$i^- = -\frac{v^-}{Z_0}$$

Combining eqns. 8.18 and 8.19 results in

$$\frac{v}{i} = Z_r = \frac{v^+ + v^-}{i^+ + i^-} = \frac{v^+ + v^-}{(v^+ - v^-)/Z_0}$$

and solving for v^- we obtain

$$v^- = \frac{Z_r - Z_0}{Z_r + Z_0} v^+ = K_r v^+$$

(8.20)

where

$$K_r = \frac{Z_r - Z_0}{Z_r + Z_0}$$

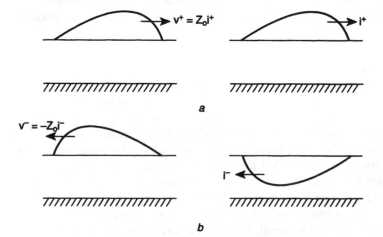

Figure 8.2   Propagation of travelling waves on assumed lossless single-phase lines

(*a*) transmitted voltage and current waveforms
(*b*) reflected voltage and current waveforms

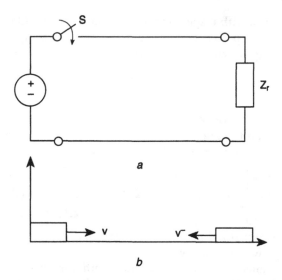

Figure 8.3   Successive wave reflection at both ends of a single-phase transmission line

(a) line circuit
(b) wave applied at the source and reflected at the load end

The constant K_r is often called the receiving end coefficient of reflection because it describes the voltage reflected at a discontinuity in terms of the incident (or forward wave component).

By combining eqns. 8.19 we similarly obtain

$$\frac{i^-}{i^+} = -\frac{v^-}{v^+} = -\frac{Z_r - Z_0}{Z_r + Z_0}$$

or

$$i^- = -K_r i^+ \qquad (8.21)$$

In a similar manner, the coefficient of reflection at the sending-end can be derived as:

$$K_s = \frac{Z_s - Z_0}{Z_s + Z_0} \qquad (8.22)$$

where Z_s is the sending end source impedance.

8.2.2 Three-phase transposed lines

When a three-phase line is perfectly transposed, the parameters are identical for each phase. In common with untransposed lines, such lines can be decomposed into three modal equivalent circuits.

Modal analysis of three-phase transmission lines [3, 4, 5]
With reference to eqns. 8.6 and 8.7, it will be seen that the elements of $P = ZY$ are independent of x. It follows that eqns. 8.6 and 8.7 can be treated as linear

differential equations with respect to voltages and currents. On the other hand it is well known in power system analysis that phase quantities can be decomposed into equivalent symmetrical components such that

$$V_p = TV_s,$$

$$I_p = TV_s$$

where V_p, I_p are the phase voltage and phase current vectors, V_s, I_s are the sequence voltage and sequence current vectors and T is the symmetrical component transformation matrix defined below:

$$V_p = \begin{bmatrix} V_a \\ V_b \\ V_c \end{bmatrix}, \ V_s = \begin{bmatrix} V_0 \\ V_1 \\ V_2 \end{bmatrix} \text{ and } T = \begin{bmatrix} 1 & 1 & 1 \\ 1 & 1\angle240° & 1\angle120° \\ 1 & 1\angle120° & 1\angle240° \end{bmatrix}$$

V_a, V_b, V_c are the voltages of phases 'a', 'b' and 'c', and V_0, V_1, V_2 are the zero, positive and negative sequence voltages.

It should be noted that, in accordance with the theory of symmetrical components, phasor values are employed throughout in the above analysis.

It is possible likewise to transform the instantaneous (or time domain) phase quantities into what are called modal components such that

$$v_p(t) = Sv_m(t)$$

$$i_p(t) = Qi_m(t)$$ (8.23)

In this case, the voltage and current vectors $v_m(t)$, $i_m(t)$ are modal component vectors defined by

$$v_m(t) = \begin{bmatrix} v^{(0)}(t) \\ v^{(1)}(t) \\ v^{(2)}(t) \end{bmatrix} \qquad i_m(t) = \begin{bmatrix} i^{(0)}(t) \\ i^{(1)}(t) \\ i^{(2)}(t) \end{bmatrix}$$

It should be noted that the superscripts (0), (1), and (2) refer to the zero, first and second modes. The matrices S and Q are the voltage and current transformation matrices, which are used to transform the phase quantities into modal components.

The elements of matrices S and Q can be determined using matrix function theory and the properties of eigenvalues and eigenvectors. Details of the analyses involved are beyond the scope of this book, but a brief outline of the method is useful.

By substituting eqns. 8.23 into eqns. 8.6 and 8.7, we obtain

$$\frac{d^2 v_m}{dx^2} = S^{-1}PS v_m$$ (8.24)

$$\frac{d^2 i_m}{dx^2} = Q^{-1}P^T Q i_m$$ (8.25)

It is necessary to choose the matrices S and Q such that $S^{-1}PS$ and $Q^{-1}P^T Q$ become diagonal. This involves finding three scalar quantities (the eigenvalues) and a column matrix X (the eigenvectors) for eqns. 8.6 and 8.7. For example, the eigenvalues and eigenvectors for eqn. 8.6 should satisfy the following equation:

$$(P - U)X = 0 \tag{8.26}$$

where U is the unit matrix.

For perfectly transposed lines, the matrices S and Q are equal to each other for some transformations, and in protection applications, three modal transformation matrices have been widely used. These are:

1. The so called Wedepohl transformation

$$S = Q = \begin{bmatrix} 1 & 1 & 1 \\ 1 & 0 & -2 \\ 1 & -1 & 1 \end{bmatrix} \tag{8.27}$$

2. The Karrenbauer transformation

$$S = Q = \begin{bmatrix} 1 & 1 & 1 \\ 1 & -2 & 1 \\ 1 & 1 & -2 \end{bmatrix} \tag{8.28}$$

3. The Clark transformation

$$S = Q = \begin{bmatrix} 1 & 1 & 0 \\ 1 & -1/2 & -3/2 \\ 1 & -1/2 & -3/2 \end{bmatrix} \tag{8.29}$$

Modal surge impedance

Because the transformation matrices S and Q are real for a transposed line, they can be applied in both the frequency and time domains. Thus, by substituting the foregoing transformations $(S = Q)$ into eqns. (8.24) and (8.25) a set of decoupled propagation characteristics can be obtained:

$$\frac{d^2 V^{(0)}}{dx^2} = (\gamma^{(0)})^2 V^{(0)}, \qquad \frac{d^2 I^{(0)}}{dx^2} = (\gamma^{(0)})^2 I^{(0)}$$

$$\frac{d^2 V^{(1)}}{dx^2} = (\gamma^{(1)})^2 V^{(1)}, \qquad \frac{d^2 I^{(1)}}{dx^2} = (\gamma^{(1)})^2 I^{(1)} \tag{8.30}$$

$$\frac{d^2 V^{(2)}}{dx^2} = (\gamma^{(2)})^2 V^{(2)}, \qquad \frac{d^2 I^{(2)}}{dx^2} = (\gamma^{(2)})^2 I^{(2)}$$

where $\gamma^{(k)}$ is the modal propagation constant of the kth mode. The zero mode is also termed the earth mode, while the second and third modes are termed 'aerial modes'.

These equations show that wave propagation in a three-phase line can be considered in terms of three separate and independent components, each with its own propagation constant $\gamma^{(k)}$ and associated surge impedance $Z_0^{(k)}$. Assuming complete transposition, the surge impedance associated with each mode can be shown to be:

$$Z_0^{(0)} = \sqrt{(Z_s + 2Z_m)/(Y_s - 2Y_m)} = \sqrt{Z_0/Y_0}$$
$$Z_0^{(1)} = Z_0^{(2)} = \sqrt{(Z_s - Z_m)/(Y_s + Y_m)} = \sqrt{Z_1/Y_1}$$

(8.31)

where $Z_0(k)$ is the k-mode surge impedance, Z_s, Y_s are the average sum of all conductor self impedances and admittances at any frequency, Z_m, Y_m are the average sum of all conductor mutual impedances and admittances at any frequency, Z_0, Y_0 are the zero-phase sequence line impedance and admittance and Z_1, Y_1 are the positive-phase sequence line impedance and admittance.

The modal propagation constants are similarly given in terms of phase sequence impedances and admittances by

$$\gamma^{(0)} = \sqrt{Z_0 Y_0}$$

(8.32)

$$\gamma^{(1)} = \gamma^{(2)} = \sqrt{Z_1 Y_1}$$

(8.33)

8.3 Superimposed quantities and their properties

When a power system is subjected to a fault condition, the total voltage and current at any point in the system can be considered as consisting of two components: one is due to the sinusoidal steady-state condition, the other due to the application of the fault. The latter component is often called the superimposed quantity and is simply equal to the change in the current and/or voltage due to a disturbance. The idea of superimposed quantities is best explained by reference to Figure 8.4. Consider point R, where the voltage $v_{Rf}(t)$ and current $i_{Rf}(t)$ are the actual quantities present (see Figure 8.4a). These quantities can be split into the sinusoidal prefault quantities $v_{RS}(t)$ and $i_{RS}(t)$ shown in Figure 8.4b, and the fault superimposed quantities $\Delta v_R(t)$ and $\Delta i_R(t)$ shown in Figure 8.4c. Thus

$$v_{Rf}(t) = v_{RS}(t) + \Delta v_R(t)$$
$$i_{Rf}(t) = i_{RS}(t) + \Delta i_R(t)$$

(8.34)

It should be noted that the sum of the steady-state and superimposed quantities is equal to the total variation at all points in the network. The superimposed network is therefore a network with all source voltages set at zero (short circuited). The superimposed fault point voltage is therefore zero for all time up to the time of the fault disturbance. The voltage of the fictitious superimposed source on the faulted phase(s) is thus equal in magnitude and opposite in sign to the prefault voltage at the fault point (i.e. $v_F(t)$).

The superimposed quantities of voltages and currents possess unique properties in relation to the location of the fault, which will be explained in the following subsections. These properties form the basis of practical travelling-wave based protection schemes.

8.3.1 Polarity of superimposed quantities versus fault location [6, 7]

We shall show in this Section that the relative polarities of the superimposed voltage and current waveforms (or the direction of the motion of their travelling waves) at the two ends of the line under consideration depend on the fault location. Consider Figure 8.5, from which it can be seen that the fault produces superimposed quantities at the relaying points R and S. If an internal fault is assumed to occur during the positive half cycle of the prefault voltage waveform, then the signs of the superimposed voltage and current would be $(-)$ and $(+)$, respectively, at both ends of the line. (Note that reference

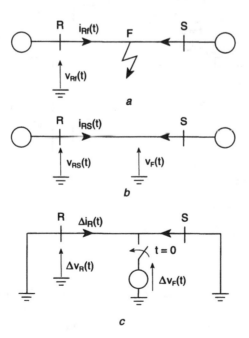

Figure 8.4 Superimposed voltage and current generated by a fault

     (*a*) actual system under a fault condition
     (*b*) the steady-state (prefault) network
     (*c*) the superimposed network

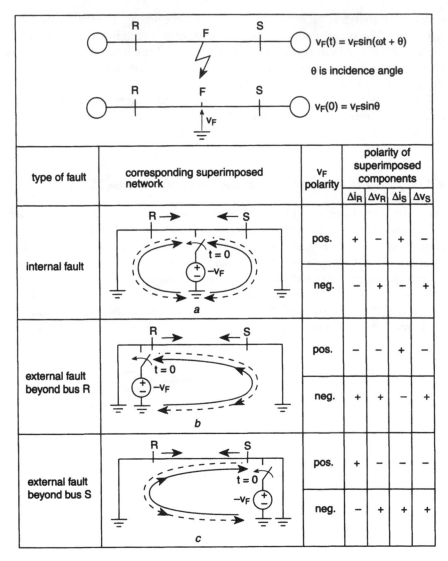

Figure 8.5   Dependency of polarities of superimposed voltage and current quantities on fault location

directions are those shown in Figure 8.4). However, if the fault occurs during the negative half cycle of the voltage waveform, the polarities of the superimposed voltage and current at the R and S ends would be $(+)$ and $(-)$ respectively. Repeating the analysis for faults at locations external to the line R–S reveals that the polarities of the superimposed voltage and current would be similar to each other at one end but are different from each other at the other end. Figure 8.5 explains in detail all possible cases.

8.3.2 Interrelation between the superimposed voltage and current quantities versus fault location [8, 9]

Consider Figure 8.6, which shows a line R–S interconnecting two systems R and S. The following analysis and associated conclusions are general, but to simplify the discussion we shall assume that

(i) the line is lossless, and therefore the surge impedance of the line (Z_0) and the wave propagation velocity c are given by

$$Z_0 = \sqrt{L/C}$$
$$c = 1/\sqrt{LC} \approx 300 \times 10^8 \text{ m/s} \tag{8.35}$$

(ii) the surge impedances looking into systems S and R are equal to Z_{0S} and Z_{0R}, respectively, and are such that

$$Z_{0R} = Z_{0S} = Z_0/3 \tag{8.36}$$

These terminating conditions are exactly analogous to a situation in which each substation terminates four lines of equal surge impedance Z_0. In the light of these assumptions, which are used in producing Figures 8.7 and 8.8, we shall examine the relationship between the superimposed voltages and currents for faults in the reverse and forward directions as seen by a relay located at end R.

Reverse faults
Let us assume that a solid fault occurs immediately behind the relay location at R. Now because the fault point coincides with the relaying point, the resulting superimposed voltage at R, $\Delta v_R(t)$, is equal and opposite to the steady-state prefault voltage waveform $v_F(t)$ at end R, i.e. $\Delta v_R(t) = -v_F(t)$. As discussed in Section 8.2.1, this causes a travelling wave to propagate towards end S at a velocity c defined by eqn. 8.35. The associated superimposed current waveform $\Delta i_R(t)$ also propagates at a velocity c and the superimposed voltage and current

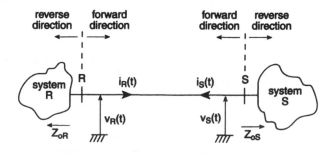

Figure 8.6   Line interconnecting two systems R and S

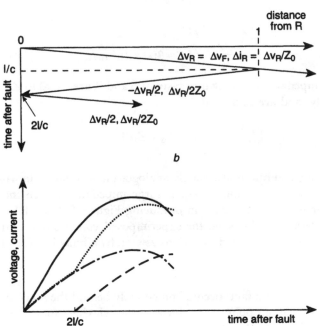

Figure 8.7   *Initial propagation of superimposed quantities following a reverse fault immediately behind the relay at R*

(*a*) superimposed fault network
(*b*) lattice diagram
(*c*) time of variation of superimposed quantities consistent with fault at the positive-to-negative zero-crossing of the prefault voltage
────── total superimposed voltage at R
$\Delta v_R(t) = \Delta v_F(t) = h(t) V_m \sin \omega_0 t$
──·── initial superimposed current $\Delta v_F(t)/Z_0$
──── total reflected superimposed current in line at R at time $2l/c$, $\Delta v_F(t-2l/c)/Z_0$
······ total superimposed current in line at R
$\Delta i_R(t) = \Delta v_F(t)/Z_0 + \Delta v_F(t-2l/c)/Z_0$

waveform $\Delta v_R(t)$ and $\Delta i_R(t)$ are interrelated according to eqn. 8.17. In the period of time from fault inception up to twice the wave transit time from end R to S ($2l/c$, l = length of line), $\Delta v_R(t) = \Delta v_R^+(t)$, $\Delta i_R(t) = \Delta i_R^+(t)$ and $\Delta v_R^-(t) =$

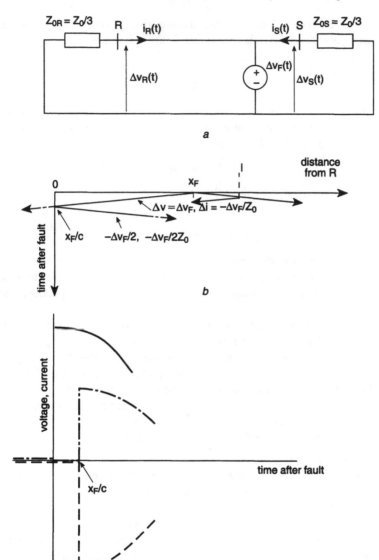

Figure 8.8 *Initial propagation of superimposed quantities following a fault in the forward direction of the relay at R*

(a) superimposed fault network
(b) lattice diagram
(c) time variation of superimposed quantities consistent with fault at negative peak of prefault voltage
———— superimposed fault voltage $v_F(t)$
———·— superimposed voltage at R
$v_R(t) = \Delta v_F(t - x_F/c)/2$
----- superimposed current at R
$i_R(t) = -3\Delta v_F(t - x_F/c)/2Z_0$

$\Delta i_R^-(t) = 0$; it therefore follows that, during this period, the superimposed voltage and current at the relaying point R are related by

$$\left. \begin{aligned} \Delta v_R(t) &= Z_0 \Delta i_R(t) \\ \Delta v_R(t) - Z_0 \Delta i_R(t) &= 0 \end{aligned} \right\} \qquad (8.37)$$

or

Because S is a point of discontinuity, part of the incident superimposed voltage and current waveforms are reflected back towards the relaying point at R, according to eqns. 8.20 and 8.21, such that

$$\begin{aligned} \Delta v_S^-(t) &= K_r \Delta v_S^+(t) \\ \Delta i_S^-(t) &= -K_r \Delta i_S^+(t) \end{aligned} \qquad (8.38)$$

where

$$K_r = (Z_{0S} - Z_0)/(Z_{0S} + Z_0)$$

and $\Delta v_S^+(t)$, $\Delta i_S^+(t)$ are the transmitted parts of superimposed voltage and current waveforms, respectively, at end S, and $\Delta v_S^-(t)$, $\Delta i_S^-(t)$ are the reflected parts of the superimposed voltage and current waveforms, respectively, at end S.

It will be evident that it takes a time of $2\,l/c$ for the waveform reflected from end S to arrive back at R after the occurrence of the fault. Figure 8.7 thus illustrates the behaviour of the system shown in Figure 8.6 due to a fault that occurs at the positive-to-negative zero crossing of prefault voltage. Figure 8.7(*b*) shows the lattice diagram of the superimposed voltages and currents at any point on the line, and Figure 8.7(*c*) shows the resulting superimposed components $\Delta v_R(t)$, $\Delta i_R(t)$ at the relaying point R. It is particularly important to note that if it is assumed that the fault occurs at time $t = 0$, then

(i) $\Delta v_R(t) = \Delta i_R(t) = 0$ for $t < 0$

(ii) Although the waveforms of Figure 8.7(*c*) are consistent with a fault at zero voltage point on wave, the relationship of eqn. 8.37 also holds for all time from fault inception up to twice the wave transit time, i.e. $2l/c$ for any arbitrary fault point superimposed voltage waveform $\Delta v_F(t)$.

(iii) The relationship described in eqn. 8.37 also holds for any value of terminating surge impedances (Z_{0R}, Z_{0S}) and line attenuation. For example, if the source surge impedance at end S were exactly equal to that of the line, there would be no reflection from S, but eqn. 8.48 still holds for all times $t < 2l/c$.

For faults within system R, there is a time delay between fault inception and the arrival of the superimposed components at the measuring point. Nevertheless, eqn. 8.37 holds for all times $t = 2l/c$ after the arrival of the superimposed quantities at R.

Forward faults

Now let us consider the case where faults occur in the forward direction with respect to the relay at end R. For example, consider a fault at a point F which is at a distance x_F from R. The resulting superimposed voltage at x_F, $\Delta v_F(t)$, is equal and opposite to the prefault voltage at that point. Again in accordance with Section 8.2.1, the application of the superimposed voltage $\Delta v_F(t)$ causes the voltage travelling wave and its associated current wave $\Delta v_F(t)/Z_0$ to propagate at a velocity c towards the point R. The superimposed voltage and superimposed current waveforms incident at R arrive at a time x_F/c and are given by

$$\Delta v_R(t) = \Delta v_R^+(t) = \Delta v_F(t - x_F/c)$$

$$\Delta i_R(t) = -\Delta i_R^+(t) = -\Delta v_F(t - x_F/c)/Z_0$$

Now the reflected components of these voltage and current waveforms for an arbitrary terminating surge impedance Z_{0R} can be found from the corresponding incident waves and the reflection coefficient:

$$\Delta v_R^-(t) = K_r\Delta v_R^+(t)$$

$$= \Delta v_F(t - x_F/c)(Z_{0R} - Z_0)/(Z_{0R} + Z_0) \tag{8.39}$$

$$\Delta i_R^-(t) = -K_r\Delta i_R^+(t)$$

$$= -\Delta v_F(t - x_F/c)(Z_{0R} - Z_0)/[(Z_{0R} + Z_0)Z_0] \tag{8.40}$$

But the total variation of superimposed voltage and current at, and immediately after, the time when the waves reach end R ($t = x_F/c$) is given by the sum of transmitted and reflected waves:

$$\Delta v_R(t) = \Delta v_R^+(t) + \Delta v_R^-(t)$$

$$= 2Z_{0R}\Delta v_F(t - x_F/c)/(Z_{0R} + Z_0) \tag{8.41}$$

$$\Delta i_R(t) = \Delta i_R^+(t) + \Delta i_R^-(t)$$

$$= -2\Delta v_F(t - x_F/c)/(Z_{0R} + Z_0) \tag{8.42}$$

Therefore, by finding the difference between eqns. 8.41 and 8.42 and their sum, we obtain

and

$$\Delta v_R(t) - Z_0\Delta i_R(t) = 2\Delta v_F(t - x_F/c)$$

$$\Delta v_R(t) + Z_0\Delta i_R(t) = 2\Delta v_F(t - x_F/c)(Z_{0R} - Z_0)/(Z_{0R} + Z_0) \tag{8.43}$$

Eqns. 8.43 show that, unlike the previously considered reverse fault, the difference $\Delta v_R(t) - Z_0\Delta i_R(t)$ becomes immediately finite on arrival of the superimposed waves at x_F/c after fault inception. This conclusion holds for faults anywhere on the line or within system S.

Figure 8.8 illustrates the behaviour of the superimposed quantities following a fault on the line R–S of the system of Figure 8.6. The fault occurs at the negative peak of the prefault voltage. It is evident from the lattice diagram and the waveforms of Figures 8.8(*b*) and (*c*), that, initially, the travelling waves of superimposed voltage and current are of opposite polarity.

However, in case of a fault at the end S, there is a time delay between fault inception and the arrival of the superimposed quantities at R. Although the waveforms shown in Figure 8.8(*c*) are consistent with a fault at peak prefault voltage, the conclusions drawn equally apply for any other arbitrary superimposed fault point voltage waveform.

Relaying signals
The difference in the behaviour of the superimposed quantities according to whether the fault is in the reverse or the forward direction can be used to determine the direction of the fault with respect to the measuring point. This is done by the formation of two signals. In terms of primary system values, the signals used at R are

$$S_{1R} = \Delta v_R(t) - R_0 \Delta i_R(t)$$
$$S_{2R} = \Delta v_R(t) + R_0 \Delta i_R(t)$$

(8.44)

The corresponding signals used at S which have the reference directions defined in Figure 8.8(*a*), are given as

$$S_{1S} = \Delta v_S(t) - R_0 \Delta i_S(t)$$
$$S_{2S} = \Delta v_S(t) + R_0 \Delta i_S(t)$$

(8.45)

where R_0 is a surge replica resistance whose value is so arranged as to match closely the line surge impedance Z_0.

By examining the signals at end R for reverse faults, with the assumption of lossless line and perfect matching of the replica resistance R_0, i.e. by substituting eqn. 8.37 into eqns. 8.44, we obtain

$$S_{1R} = 0$$
$$S_{2R} = 2\Delta v_R(t)$$

(8.46)

Figure 8.9(*a*) shows how the signals vary following the reverse fault condition considered previously. It is important to note that S_{2R} becomes finite before S_{1R}. Figure 8.9(*a*) also shows the sequence in which the signals S_{1R} and S_{2R} rise above a given threshold value $|V_s|$, (S_{2R} followed by S_{1R}), at times $T_1 < T_2$.

Similarly, the signals S_{1R} and S_{2R} for the forward fault described previously can be found by substituting eqn. 8.43 into eqn. 8.44. By assuming perfect matching of the replica resistance R_0 to the surge impedance Z_0, these signals would be given as

$$S_{1R} = 2\Delta v_F(t - x_F/c)$$
$$S_{2R} = 2\Delta v_F(t - x_F/c)(Z_{0R} - Z_0)/(Z_{0R} + Z_0)$$

(8.47)

Figure 8.9(*b*) illustrates the behaviour of the signals S_{1R}, S_{2R} due to the forward fault condition on the system of Figure 8.8. It can be seen that the setting levels V_s and $-V_s$ are exceeded almost simultaneously ($T_1 \approx T_2$) owing to the very

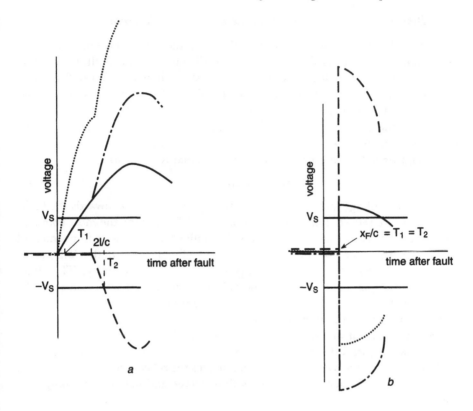

Figure 8.9 *Variation of relaying signals at R following faults on the system of Figure 8.6*

       (*a*) reverse fault at R
       (*b*) fault on line R–S
       ———— $\Delta v_R(t)$
       —·—· $R_0 \Delta i_R(t)$
       – – – – $S_{1R} = \Delta v_R(t) - R_0 \Delta i_R(t)$
       ······· $S_{2R} = \Delta v_R(t) + R_0 \Delta i_R(t)$

rapid change in the superimposed components associated with the assumed fault at peak prefault voltage.

When a three-phase line is considered, the phase variations of currents and voltages are decomposed into modal variations, and the relaying signals can therefore be expressed in terms of aerial-mode surge impedances $Z_0^{(1)}$ and $Z_0^{(2)}$, and the earth-mode surge impedance $Z_0^{(0)}$ (see Section 8.2.2). Assuming a lossless three-phase line with complete transposition and perfect matching of the replica surge impedances $R_0^{(k)}$, $k = 0$, 1 and 2 to the corresponding surge impedance $Z_0^{(k)}$, $k = 0$, 1 and 2, the modal relaying signals at the end R of the line are given by

$$S_{1R}^{(k)} = \Delta v_R^{(k)}(t) - Z_0^{(k)} \Delta i_R^{(k)}(t)$$
$$S_{2R}^{(k)} = \Delta v_R^{(k)}(t) + Z_0^{(k)} \Delta i_R^{(k)}(t) \qquad k = 0, 1, 2 \qquad (8.48)$$

8.3.3 Behaviour of relaying signals at the relay and fault locations [10]

The way relaying signals S_1 and S_2 behave at the relay and fault points of discontinuity can be used to estimate the distance of the fault to a relay. Consider a forward fault on the line of the system shown in Figure 8.10. The superimposed voltage at the fault point F is $\Delta v_F(t)$ and its associated superimposed current $\Delta v_F(t)/Z_0$ propagates towards end R.

Thus, if the relaying signal S_{1R} (see eqn. 8.44) is formed from the above voltage and current before and after they are reflected at the relay busbar we would find its value does not change, and it remains equal to

$$S_{1R} = 2\Delta v_F(t - x_F/c) \tag{8.49}$$

The reflected voltage and current waves then propagate back towards the fault, where they are reflected back again towards end R. Assume the most common case, where the fault resistance R_f is much smaller than the surge impedance of the line. Under this condition the coefficient of reflection at the fault point $K_f = (R_f - Z_0)/(R_f + Z_0)$, would be approximately equal to unity. Therefore, according to eqns. 8.20 and 8.21, the voltage and current waveforms will be reflected at the fault with almost the same magnitudes. However, the reflected voltage would have the same polarity as the incident voltage waveform, while the reflected current would have opposite polarity. If S_{1R} is recalculated from the reflected waves, its value would be zero. The important conclusion from this analysis is that the relaying signal S_{1R} that leaves the relay point toward the fault would be reflected back to the relay, with its shape and magnitude changed at the fault point.

A similar analysis can be carried out with regard to S_{2R}, which reveals that it is changed at the relaying point, in exactly the same manner as the signal S_{1R}. It follows that, if the time between like changes in S_{1R} and S_{2R} is equal to τ, the distance to the fault x_F would be

$$x_F = \frac{\tau/2}{c} \tag{8.50}$$

Figure 8.10 shows the response of a single-phase lossless line to a 1 p.u. voltage step function applied at the fault point, for values of fault resistance R_f of zero and of 25% of the characteristic impedance of the line. The coefficients of reflection are calculated on the basis that inductive sources are connected at buses 1 and 2, having short-circuit capacities of 35 000 MVA and 10 000 MVA, respectively.

8.3.4 Superimposed component elliptical trajectories [11]

For the following discussion, let us consider the system shown in Figure 8.11. If we assume the prefault voltage at the fault point F is $v_F(t) = V\sin(\omega_0 t + \theta)$, where V and θ are the peak voltage and the inception angle, respectively, then the superimposed voltage and current waveforms $\Delta v_R(t)$ and $\Delta i_R(t)$ at end R can be calculated as explained below.

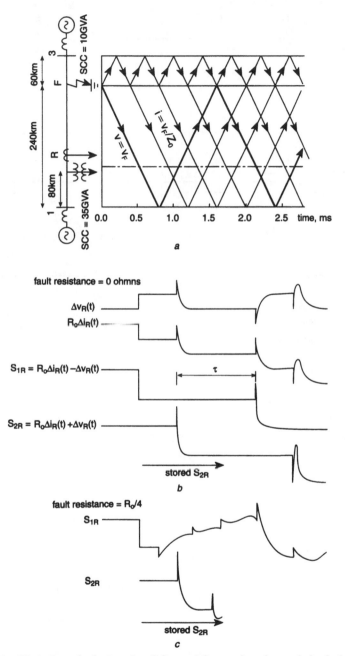

Figure 8.10   Variation of relaying signal S_{1R} and S_{2R} at the relay and the fault points

(a) system under fault condition with the corresponding lattice diagram
(b) voltage and current waveforms due to a zero resistance fault and the corresponding signals S_{1R} and S_{2R} with zero fault resistance
(c) the signals S_{1R} and S_{2R} due to the same fault but with fault resistance equal to 25% of surge impedance

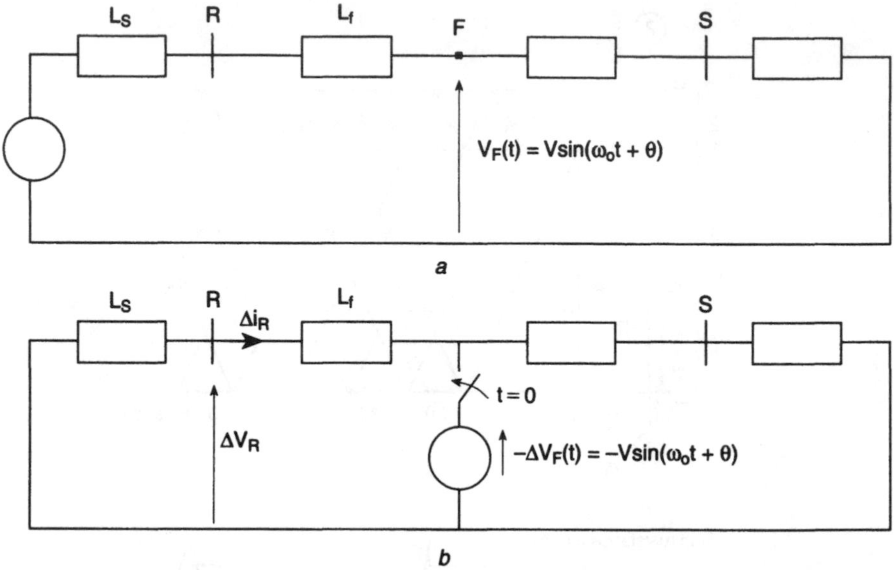

Figure 8.11   Representation of single-phase system used to explain the elliptical trajectory based technique

(*a*) steady-state network
(*b*) superimposed network

With reference to Figure 8.11(*b*), the superimposed voltage $-V\sin(\omega_0 t + \theta)$ causes a superimposed current $\Delta i_R(t)$ to flow according to the following differential equations:

$$(L_s + L_f)\frac{d\Delta i_R(t)}{dt} = \Delta v_F(t)$$

$$= -V\sin(\omega_0 t + \theta) \qquad (8.51)$$

By solving eqn. 8.51, we obtain

$$\Delta i_R(t) = \frac{V}{\omega_0(L_s + L_f)}(\cos(\omega_0 t + \theta) - \cos\theta)$$

or

$$\Delta i_R(t) = \Delta I(\cos(\omega_0 t + \theta) - \cos\theta) \qquad (8.52)$$

where

$$\Delta I = \frac{V}{\omega_0(L_s + L_f)}$$

Again with reference to Figure 8.11(b), the superimposed voltage $\Delta v_R(t)$ can be determined from the superimposed current $\Delta i_R(t)$:

$$\Delta v_R(t) = -L_s \frac{d\Delta i_R(t)}{dt} \tag{8.53}$$

By substituting eqn. 8.52 into eqn. 8.53 we obtain

$$\Delta v_R(t) = \omega_0 L_s \Delta I \, \sin(\omega_0 t + \theta)$$

or

$$\Delta v_R(t) = \Delta V \sin(\omega_0 t + \theta) \tag{8.54}$$

where

$$\Delta V = \omega_0 L_s \Delta I = \frac{X_s V}{X_s + X_f}$$

and $X_s = \omega_0 L_s$ is the reactance of the source, and $X_f = \omega_0 L_f$ is the reactance of the line to the fault.

Multiplying eqn. 8.52 by a mimic resistance R and combining the resulting equation with eqn. 8.54 results in

$$\frac{\Delta v_R^2(t)}{(\Delta V)^2} + \frac{(R\Delta i_R(t) - R\Delta I \cos\theta)^2}{(R\Delta I)^2} = 1 \tag{8.55}$$

By examining eqns. 8.52 to eqn. 8.55, the following conclusions can be drawn:

(i)   Eqn. 8.52 shows that the superimposed current component $\Delta i_R(t)$ contains a DC offset equal to $\Delta I \cos\theta$, which depends on the fault inception angle θ.

(ii)   Inclusion of resistance in the faulted loop will, of course, cause a decay of the DC offset. On the other hand, such resistance will slightly reduce the magnitudes of ΔV and ΔI and introduce an additional phase shift between the fundamental frequency components of the superimposed quantities.

(iii)   Eqn. 8.55 represents an ellipse in the superimposed plane (i.e. the $R\Delta i_R(t) - \Delta v_R(t)$ plane), with its centre being shifted in the direction of $R\Delta i_R(t)$ by a constant value of $R\Delta I \cos\theta$. This ellipse is known as the fault trajectory.

Figure 8.12(a) shows the steady-state voltage $v_{SR}(t)$, the superimposed voltage Δv_R and the superimposed current Δi_R due to the fault under consideration. Figure 8.12(b) shows the fault trajectory. It can be seen that immediately after the fault, Δv_R will have a finite value, while Δi_R (under the assumption of a lossless line) changes gradually from zero. Therefore, the trajectory initially jumps from the origin of the plane to a point somewhere along the Δv_R axis. It will then proceed along the elliptical path of Figure 8.12(b). According to eqn. 8.54, the magnitude of initial jump Δv_R is a maximum when the fault inception angle θ is equal to $\pm n\pi/2$, i.e. the prefault voltage is maximum, while Δv_R is zero for a zero fault inception angle.

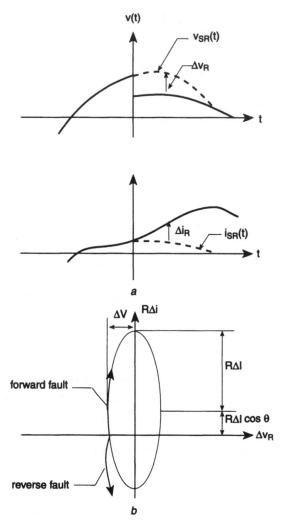

Figure 8.12 (*a*) Behaviour of voltage and current waveforms under fault
conditions
(*b*) corresponding fault trajectory

8.4 Bergeron's equations [12, 13]

8.4.1 Single-phase lines

To simplify the analysis once more, a lossless single-phase line will be
considered so that, by combining eqns. 8.14 and 8.15, we obtain

$$i(x,\ t) + \frac{1}{Z_0}\ v(x,\ t) = 2F_1\!\left(t - \frac{x}{c}\right)\!\bigg/ Z_0 \qquad (8.56)$$

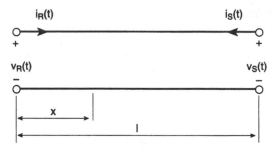

Figure 8.13   Single-phase line used to explain Bergeron's equations

If we consider the transmission line shown in Figure 8.13, eqn. 8.56 can be applied to the R and S ends, where $x = 0$ and $x = l$, respectively. The following equations are then obtained:

$$i(0, t) + \frac{1}{Z_0} v(0, t) = 2F_1(t)/Z_0 \tag{8.57}$$

$$i(l, t) + \frac{1}{Z_0} v(l, t) = 2F_1\left(t - \frac{l}{c}\right) \Big/ Z_0 \tag{8.58}$$

By renaming $i(0, t)$, $v(0, t)$ as $i_R(t)$ and $v_R(t)$, $i(l, t)$ and $v(l, t)$ as $-i_S(t)$ and $v_S(t)$, substituting t by $(t - l/c)$ in eqn. 8.57, and combining the resulting equation with eqn. 8.58, we obtain Bergeron's equation:

$$i_R(t - \tau) + \frac{1}{Z_0} v_R(t - \tau) = -i_S(t) + \frac{1}{Z_0} v_S(t) \tag{8.59}$$

where $\tau = l/c$ is the surge transit time.

The right-hand side of eqn. 8.59 is twice the forward component of the current originating from end R of the line at time $(t - \tau)$ (see eqn. 8.57). Similarly, the left-hand side of the equation equals twice the same forward current component originating from end R and arriving at S at time t, i.e. after a time delay of τ where $\tau = l/c$, which is the travelling time of the forward current component over a line of length l.

Eqn. 8.59 therefore means simply that the forward current component originating from one end of a healthy line arrives at the other end after a time delay equal to the surge time of the line.

A further 'Bergeron's equation' can be obtained by eliminating $F_2(t + x/c)$ from eqns. 8.14 and 8.15, and this in turn results in

$$-i_R(t) + \frac{1}{Z_0} v_R(t) = i_S(t - \tau) + \frac{1}{Z_0} v_S(t - \tau) \tag{8.60}$$

The physical interpretation of this equation is that a backward component originating from one end of a healthy line arrives at the other end after a time delay equal to the line surge time.

8.4.2 Three-phase lines

To find Bergeron's equations for a three-phase line, the phase variations of currents and voltages are first decomposed into modal components. Each modal equation is then treated in exactly the same manner as for a single-phase line. If the line is assumed to be lossless and ideally transposed, eqns. 8.5 can be reduced in terms of line capacitances and inductances as follows:

$$\frac{\partial v}{\partial x} = Li$$

$$\frac{\partial i}{\partial x} = Ci \tag{8.61}$$

When these equations are decomposed into their modal components as explained in Section 8.2.2, we obtain

$$\frac{\partial v_m}{\partial x} = L_m i_m$$

$$\frac{\partial i_m}{\partial x} = C_m v_m \tag{8.62}$$

where

$$L_m = \begin{bmatrix} L^{(0)} & 0 & 0 \\ 0 & L^{(1)} & 0 \\ 0 & 0 & L^{(2)} \end{bmatrix} \qquad C_m = \begin{bmatrix} C^{(0)} & 0 & 0 \\ 0 & C^{(1)} & 0 \\ 0 & 0 & C^{(2)} \end{bmatrix}$$

In fact, eqns. 8.62 represent independent sets of modal equations. Therefore, the derivation procedure outlined previously for the single-phase line is applicable to each of these modal equations. Thus, for a three-phase line, Bergeron's equation can be derived from eqns. 8.59 and 8.60 as follows:

$$i_R^{(k)}(t - \tau^{(k)}) + \frac{1}{Z_0^{(k)}} v_R^{(k)}(t - \tau^{(k)})$$

$$= -i_S^{(k)}(t) + \frac{1}{Z_0^{(k)}} v_S^{(k)}(t)$$

$$-i_R^{(k)}(t) + \frac{1}{Z_0^{(k)}} v_R^{(k)}(t) \tag{8.63}$$

$$= i_S^{(k)}(t - \tau^{(k)}) + \frac{1}{Z_0^{(k)}} v_S^{(k)}(t - \tau^{(k)})$$

where $\tau^{(k)}$ is the k mode surge travelling time, $Z_0^{(k)}$ is the k mode surge impedance, $\tau^{(k)} = l\sqrt{L^{(k)}C^{(k)}}$ and $Z_0^{(k)} = \sqrt{L^{(k)}/C^{(k)}}$.

8.5 Discriminant functions [5, 14, 15]

It is possible to differentiate between internal and external (or forward and reverse) faults with respect to a power line terminal by employing suitable discriminative functions. The basis of this technique is given in this section.

8.5.1 Single-phase lines

Let us again consider a faulted lossless single-phase line (Figure 8.14(a)). As established previously, the fault can be simulated as a sudden application of a superimposed voltage $\Delta v_F(t)$ at the fault point F, where $\Delta v_F(t)$ is equal and opposite to the prefault steady-state voltage at point F. This causes a superimposed voltage $\Delta v_R(t)$ and current $\Delta i_R(t)$ to appear at end R.

By applying Bergeron's eqn. 8.59 to the section R–F of the line, and substituting Δv_F, Δi_F, Δv_R, Δi_R and $(t+\tau)$ for v_R, i_R, v_S, i_S and t, we obtain

$$\Delta v_R(t+\tau) - Z_0\Delta i_R(t+\tau) = \Delta v_F(t) + Z_0\Delta i_F(t) \tag{8.64}$$

Now let us assume the prefault voltage at the fault point F, immediately before the occurrence of the fault, is $v_F(t)$, given by

$$v_F(t) = V\sin(\omega_0 t + \theta)$$

Therefore

$$\Delta V_F(t) = -V_S\sin(\omega_0 t + \theta) \tag{8.65}$$

where V and θ are as defined previously.

Since the backward or reflected travelling wave b shown in Fig 8.14(d) is initially zero, the current change $\Delta i_F(t)$ at the fault point during the period between the arrival of the wave at end R and its return back to the fault point F will be given as follows (see eqns. 8.14 and 8.15):

$$\Delta i_F(t) = \Delta i_F^+(t) = \frac{\Delta v_F^+(t)}{Z_0}$$

$$= -V\sin(\omega_0 t + \theta)/Z_0 \qquad \tau \leqslant t \leqslant 2\tau \tag{8.66}$$

The forward wave f, i.e. $\Delta v^+(t)$ and/or $\Delta i^+(t)$, which propagates towards end R, and the backward (or reflected) wave b, which is reflected back towards the fault point F, are shown in Figure 8.14(d).

By substituting $\Delta v_F(t)$ and $\Delta i_F(t)$, given by eqns. 8.65 and 8.66, respectively, into eqn. 8.64 we obtain

$$\Delta v_R(t+\tau) - Z_0\Delta i_R(t+\tau) = -2V\sin(\omega_0 t + \theta) \qquad \tau \leqslant t \leqslant 2\tau \tag{8.67}$$

It can be seen that eqn. 8.67, which describes the wave characteristics as seen at end R, depends on the fault inception angle θ. This dependency can be removed

by taking the derivative with respect to time of the wave characteristic eqn.
8.67, which in turn gives

$$\frac{d}{dt}\left(\Delta v_R(t) - Z_0\Delta i_R(t)\right) = -2\omega_0 V \cos(\omega_0 t + \theta')$$ (8.68)

where $\theta' = \theta - \omega_0\tau$.

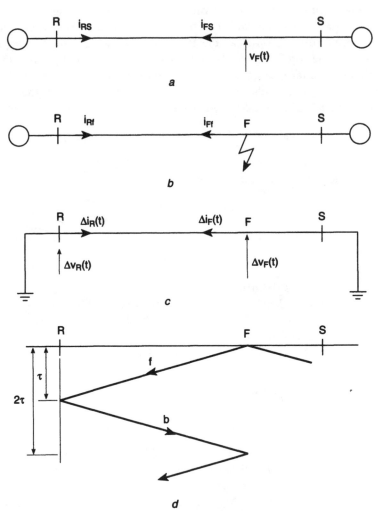

Figure 8.14   Single-phase system used to explain the forward discriminant function

(*a*) system under steady-state conditions
(*b*) system under a forward fault condition with respect to a relay at end R
(*c*) superimposed network
(*d*) the corresponding forward and backward waveforms, *f* and *b* respectively

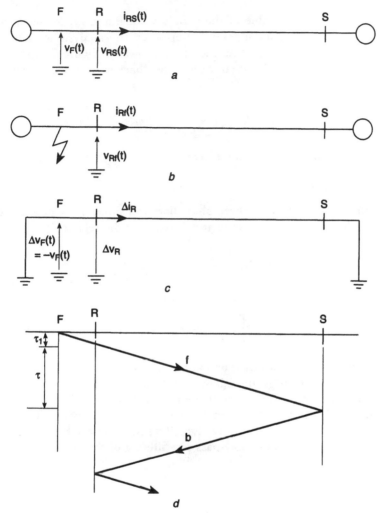

Figure 8.15 Single-phase system used to explain the backward discriminant function

    (*a*) steady-state system
    (*b*) system under a reverse (or backward) fault condition with respect to a relay at end R
    (*c*) the corresponding superimposed network
    (*d*) the corresponding forward and backward travelling waveforms, *f* and *b* respectively

This equation can be combined with eqn. 8.67 to give eqn. 8.69, which describes the wave characteristic as seen by a relay located at end R due to a fault on the line:

$$D_F = (\Delta v_R(t) - Z_0\Delta i_R(t))^2 + \frac{1}{\omega_0^2}\left[\frac{d}{dt}(\Delta v_R(t) - Z_0\Delta i_R(t))\right]^2$$

$$= 4V^2 = 8\ V_{RMS}^2 \qquad \text{for } \tau \leqslant t \leqslant 2\tau \qquad (8.69)$$

where V_{RMS} is the RMS value of the supply voltage and D_F is known as the 'forward wave discriminant function'. Repeating the analysis for faults that occur behind the relay located at end R, as shown in Figure 8.15, results in another function, which is often called the 'backward wave discriminant function' D_B

$$D_B = (\Delta v_R(t) + Z_0 \Delta i_R(t))^2 + \frac{1}{\omega_0^2}\left[\frac{d}{dt}(\Delta v_R(t) + Z_0 \Delta i_R(t))\right]^2$$

$$= 8\, V_{RMS}^2 \qquad \text{for } \tau_1 \leqslant t \leqslant \tau_1 + 2\tau \tag{8.70}$$

8.5.2 Three-phase lines

The discriminant function for three-phase lines can similarly be found in terms of the superimposed modal voltage and currents at the relaying points:

$$D_F^{(k)} = (\Delta v_R^{(k)} - Z_0^{(k)}\Delta i_R^{(k)})^2 + 1/\omega_0^2\left\{\frac{d}{dt}(\Delta v_R^{(k)} - Z_0^{(k)}\Delta i_R^{(k)})\right\}^2 \qquad k = 0, 1, 2 \tag{8.71}$$

for the modal (k) forward discriminant function, and

$$D_B^{(k)} = (\Delta v_R^{(k)} + Z_0^{(k)}\Delta i_R^{(k)})^2 + 1/\omega_0^2\left[\frac{d}{dt}(\Delta v_R^{(k)} + Z_0^{(k)}\Delta i_R^{(k)})\right]^2 \qquad k = 0, 1, 2 \tag{8.72}$$

for the modal (k) backward discriminant function.

$Z_0^{(k)}$ is the mode k surge impedance, and $\Delta v_R^{(k)}$ and $\Delta i_R^{(k)}$ are the modal k superimposed voltage and current at the relay location end R.

In practice, algorithms based on this technique can be very sensitive to noise because of the need to use time derivatives. This problem can be ameliorated in some applications by suitable low-pass pre-filtering of the measurements.

8.6 References

1 BEWLEY, L.V.: 'Travelling waves on transmission systems (2nd edition)' (John Wiley, 1951)
2 BICKFORD, J.P., MULLINEUX, N., and REED, J.R.: 'Computation of power system transients' (IEE Monograph Ser. 18, Peter Peregrinus, 1976)
3 WEDEPOHL, L.M.: 'Application of matrix methods to the solution of travelling-wave phenomena in polyphase systems', *Proc. IEE*, 1963, **110**, pp. 2200–2212
4 DOMMEL, H.W.: 'Digital computer solution of electromagnetic transients in single and multiphase networks', *IEEE Trans.* 1969, **PAS-88**, pp. 388–396
5 DOMMEL, H.W., and MICHELS, J.M.: 'High speed relaying using travelling wave transient analysis', PES Winter Meeting, 1978, Paper A78 214–9
6 CHAMIA, M., and LIBERMAN, S.: 'Ultra high speed relay for EHV/UHV transmission lines—development, design and application', *IEEE Trans.* 1978, **PAS-97**, pp. 2104–2116
7 YEE, M.T., and ESZTERGALYOS, J.: 'Ultra high speed relay for EHV/UHV transmission lines- installation-staged fault tests and operational experience', *ibid*, pp. 1814–1825

8 JOHNS, A.T.: 'New ultra-high-speed directional comparison technique for the protection of E.H.V. transmission lines', *IEE Proc. C*, 1980, **127**, pp. 228–239

9 JOHNS, A.T., MARTIN, M.A., BARKER, A., WALKER, E.P., and CROSSLEY, P.A.: 'A new approach to EHV directional comparison protection using digital signal processing techniques', *IEEE Trans.* 1986, **PWRD-1**, pp. 24–34

10 CROSSLEY, P.A., and McLAREN, P.G.: 'Distance protection based on travelling waves', *IEEE Trans.* 1983, **PAS-102**, pp. 2971–2983

11 VITINS, M.: 'A fundamental concept for high speed relaying', *IEEE Trans.* 1981, **PAS-100**, pp. 163–173

12 TAKAGI, T. *et al.*: 'Fault protection based on travelling wave theory, Part I: theory', PES Summer Meeting, Mexico, 1977, Paper A77 750–753

13 TAKAGI, T., *et al.*: 'Fault protection based on travelling-wave theory, Part II: sensitivity analysis and laboratory test', PES Winter Meeting, 1978, Paper A78 220–226

14 MANSOUR, M.M., and SWIFT, G.W.: 'A multi-microprocessor based travelling wave relay—theory and realization', *IEEE Trans.* 1986, **PWRD-1**, pp. 273–279

15 MANSOUR, M.M., and SWIFT, G.W.: 'Design and testing of a multi-microprocessor travelling wave relay', *ibid*, pp. 74–82

Chapter 9

Travelling-wave protective schemes

9.1 Introduction

The previous Chapter laid down the foundation of travelling-wave protective methods. This Chapter is designed to extend knowledge about travelling-wave techniques by presenting the underlying principles of some specific implementations. Information on further development of specific products is available from manufacturers' literature.

9.2 Bergeron's-equation based scheme [1, 2]

9.2.1 Principles of internal fault detection

The principle of detecting an internal fault on a transmission line can best be explained using a single-phase lossless line model, such as that shown in Figure 9.1(a), whose voltages and currents at the two terminals of the line are related by Bergeron's eqns. 8.59 and 8.60; these are rewritten below:

$$i_R(t-\tau) + \frac{1}{Z_0} v_R(t-\tau) = -i_S(t) + \frac{1}{Z_0} v_S(t) \qquad (9.1)$$

$$-i_R(t) + \frac{1}{Z_0} v_R(t) = i_S(t-\tau) + \frac{1}{Z_0} v_S(t-\tau) \qquad (9.2)$$

As explained in Chapter 8, eqn. 9.1 implies that a travelling wave leaving end R will arrive at end S after a time delay of τ, while eqn. 9.2 implies that a travelling wave leaving S will arrive at R after a similar time delay. For faults external to the line, eqns. 9.1 and 9.2 are satisfied. However, when a fault occurs

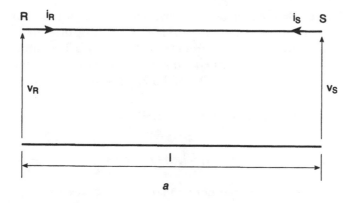

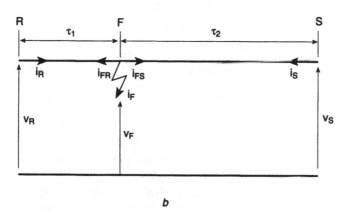

Figure 9.1 *Single-phase line*

(*a*) with no fault
(*b*) with internal fault

on the line, they are not satisfied, because a discontinuity occurs between each end. Thus, the equations can be written in the form of eqn. 9.3, in which

$\varepsilon_1(t) = 0$, $\varepsilon_2(t) = 0$ for an external fault and

$\varepsilon_1(t) \neq 0$, $\varepsilon_2(t) \neq 0$ for an internal fault,

where

$$\varepsilon_1(t) \triangleq i_R(t-\tau) + i_S(t) + \frac{1}{Z_0} [v_R(t-\tau) - v_S(t)]$$

$$\varepsilon_2(t) \triangleq i_S(t-\tau) + i_R(t) + \frac{1}{Z_0} [v_S(t-\tau) - v_R(t)]$$

(9.3)

It is possible to relate either of the above functions to a fault current $i_F(t)$. Consider Figure 9.1(*b*), where an internal fault is assumed to occur at a point F. If τ_1 is the surge travelling time from R to F and τ_2 is the surge travelling timefrom F to S, then Bergeron's equation for the sections of the line S–F and F–R can be obtained using eqns. 9.1 and 9.2, respectively, such that

$$i_{FS}(t-\tau_2) + \frac{1}{Z_0} v_F(t-\tau_2) = -i_S(t) + \frac{1}{Z_0} v_S(t)$$

$$-i_{FR}(t-\tau_2) + \frac{1}{Z_0} v_F(t-\tau_2) = i_R(t-\tau_1-\tau_2) + \frac{1}{Z_0} v_R(t-\tau_1-\tau_2)$$

Combining these equations and noting that $\tau_1 + \tau_2 = \tau$, results in

$$i_R(t-\tau) + i_S(t) + \frac{1}{Z_0}(v_R(t-\tau) - v_S(t)) = -(i_{FS}(t-\tau_2) + i_{FR}(t-\tau_2))$$

Now it can be seen from Figure 9.1 and eqn. 9.3 that

$$\varepsilon_1(t) = i_F(t-\tau_2) \tag{9.4}$$

Similarly it can be shown that

$$\varepsilon_2(t) = i_F(t-\tau_1) \tag{9.5}$$

It will be evident that the functions $\varepsilon_1(t)$ and $\varepsilon_2(t)$ effectively describe the current in the fault path in terms of the voltages and currents at the line ends. A nonzero value for $\varepsilon_1(t)$ and $\varepsilon_2(t)$ effectively signifies that a fault exists between the measuring points, a fact that can be used to generate a trip signal. The formation of the signals $\varepsilon_1(t)$ and $\varepsilon_2(t)$ require the transmission of the voltage and current waveforms between the line ends to form, in effect, a unit protection scheme. This technique has recently been implemented for plain feeders [3, 4] and has been extended to Teed power circuits [5]. Further details of the implementation are given in Chapter 11.

Three-phase lines
The concept of detecting internal faults using Bergeron's equation can easily be extended to three-phase lines. Consider the three-phase line shown in Figure 9.2. Assume the line is transposed, lossless and has a series per unit length inductance matrix $\mathbf{L}$ and a shunt capacitance matrix $\mathbf{C}$. The line can be decomposed into three equivalent single phase lines using eqns. 8.23, such that

$$v_m(t) = \mathbf{S}^{-1} v_p(t)$$

$$i_m(t) = \mathbf{Q}^{-1} i_p(t) \tag{9.6}$$

As previously explained, the voltage vector $v_m(t)$ defines three separate modal or component voltages, which can be derived from the actual phase voltages described by the vector $v_p(t)$. The component currents $i_m(t)$ can be similarly described in terms of the phase currents. The single-phase theory developed above can then be applied to each component or mode voltage and current thereby derived, and the previously derived discontinuity functions of eqn. 9.3 can be written for each mode component. For the kth mode eqn. 9.3 is thus written as

$$\varepsilon_1^{(k)}(t) = i_R^{(k)}(t-\tau^{(k)}) + i_S^{(k)}(t) + \frac{1}{Z_0^{(k)}} \left[v_R^{(k)}(t-\tau^{(k)}) - v_S^{(k)}(t) \right]$$

$$\varepsilon_2^{(k)}(t) = i_S^{(k)}(t-\tau^{(k)}) + i_R^{(k)}(t) + \frac{1}{Z_0^{(k)}} \left[v_S^{(k)}(t-\tau^{(k)}) - v_R^{(k)}(t) \right]$$

(9.7)

where $Z_0^{(k)}$ is the kth mode surge impedance and $\tau^{(k)}$ is the kth mode surge travelling wave time, where $k = 0, 1, 2$ for a three-phase line.

Thus the fault detection criteria in a three-phase line can be written as follows:

• If $\varepsilon^{(0)}(t) = 0$ and $\varepsilon^{(1)}(t) = 0$ and $\varepsilon^{(2)}(t) = 0$, then the line is healthy
• If $\varepsilon^{(0)}(t) \neq 0$ or $\varepsilon^{(1)}(t) \neq 0$ or $\varepsilon^{(2)}(t) \neq 0$, then an internal fault has occurred.

The physical meaning of $\varepsilon_1^{(k)}$ and $\varepsilon_2^{(k)}$ is that they describe the modal component values of the current in the fault path, i.e.

$$\varepsilon_1^{(k)}(t) = i_F^{(k)}(t-\tau_2^{(k)}), \ k = 0, 1 \text{ and } 2$$

$$\varepsilon_2^{(k)}(t) = i_F^{(k)}(t-\tau_1^{(k)}), \ k = 0, 1 \text{ and } 2$$

(9.8)

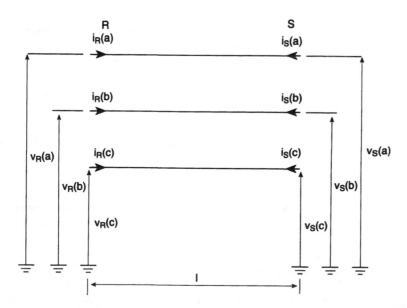

Figure 9.2 Distributed three-phase line model

where $i_F^{(k)}(t)$ is the kth modal component of fault path current, $\tau_1^{(k)}$ is the kth modal surge travelling wave time from the point F to the terminal S and $\tau_2^{(k)}$ is the kth modal surge travelling wave time from the fault point F to the terminal R.

9.3 Ultra-high-speed polarity comparison scheme [6, 7]

An ultra-high-speed relay suitable for EHV/UHV applications was developed by Asea, as a part of a joint American–Swedish research programme. The first relay developed was implemented using analogue circuits and was designated 'RALDA'; it formed the basis of a number of digital protection equipments which have emerged since the original research was performed in the late 1970s.

9.3.1 Basic operating principle

The operating principle of the scheme is based on the relative polarities of the superimposed quantities. As explained in Section 8.3.1, for internal faults the superimposed current and voltage have different polarities at both ends of the protected line. Conversely, for external faults the polarities are different only at one end of the line.

9.3.2 Description of typical implementation [6]

Figure 9.3 illustrates schematically the basic arrangement. Basically, this figure shows a transmission line which is protected by two RALDA type relays, each installed at one end of the line. The relays communicate with each other by means of a communication channel linking the two relays. This enables them to exchange information about the relative polarities of the superimposed quantities at each end, which in turn is used to decide whether the fault is internal or external to the line. The overall arrangement forms a directional comparison scheme, in which, if both relays detect a fault in a direction looking into the line, a trip signal is generated. Any one relay identifying a fault looking into either source thus indicates that the fault is external to the protected line, and in this case breaker tripping is inhibited. The actual measurements are based on information derived at single measuring points, i.e. at each end separately, and the scheme is thus a non-unit measuring technique which is effectively unitised by employing a directional signalling (binary) channel.

Each relay consists of an analogue interface, steady-state frequency suppressors, amplifiers, directional detectors, logic and drivers and an output interface. The analogue interface accepts input signals from the current and voltage transformers and provides the required galvanic isolation and surge immunity.

The steady-state frequency suppressors are designed to extract the superimposed current and voltage components caused by the fault. This is achieved by suppressing prefault components contained in the waveforms at the output of the analogue interface. The signals are then connected to amplifiers, which in effect control the basic sensitivity and are designed to achieve a minimum or

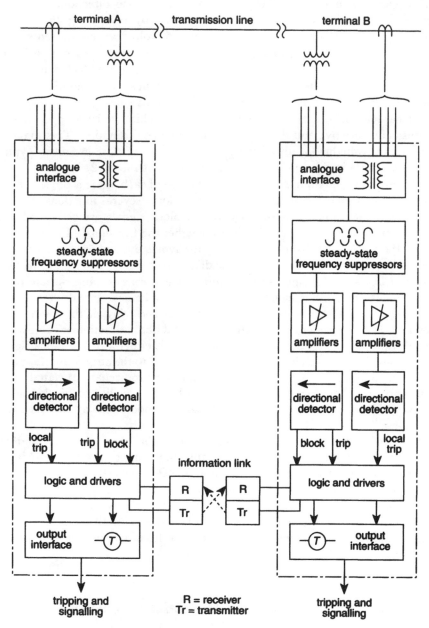

Figure 9.3   Block schematic diagram for RALDA system [6]

maximum reach along the line, depending on whether an over-reach or under-reach operation is desired. The signals are then fed to directional detectors where the direction of the fault is determined by comparing the relative polarities, i.e. the signs of the superimposed voltages and currents. The detectors then initiate a tripping direction signal T if the signs of i and v are opposite to each other. The other set of detectors seen on the block diagram are used in a parallel mode to enhance the dependability and speed of operation. Each of these units has an under-reach setting, which enables direct tripping of the local circuit breaker for close-up faults. This mode of operation is intrinsically selective and does not require time co-ordination. This makes it particularly valuable for ultra-high-speed clearance of close-in faults. Typical operating times in this mode of operation can be as low as 2–3 ms.

The output signal from the directional element of each phase is processed in a logic circuit, which has been arranged to perform several functions. First, it determines the sequence in which trip and block signals are generated and, directly on fault inception, it establishes whether the trip signal (T) was sensed before the block signal (B) or vice versa. This avoids the confusion that could be caused by multiple reflections of the incident waves. It also determines whether the fault is a single- or multi-phase fault and performs the appropriate phase selection as required in single-phase tripping schemes. In addition, it co-ordinates the local relay decision with the remote end relay via the communication link. Because each relay terminal has a local phase selection capability, the signal transmitted over the communication link is either a trip or block signal.

The signals from each logic circuit are connected to the output interface unit, which includes solid state tripping relays and electromechanical signal relays with corresponding flag indicators. It also has the facility to display signal tripping, blocking, phase selection and mode of operation, as well as alarm functions.

To provide full galvanic isolation between the relay circuitry and the auxiliary power source and to suppress any surges imposed on the DC input, the auxiliary power is supplied through a DC/DC converter.

As far as the design of the communication link is concerned, the most important factor to be taken into account is the speed. This is determined by the clearing time requirements for remote end faults, and an all-inclusive relaying time of approximately 8 ms for remote-end faults has been specified for this scheme.

9.4 Ultra-high-speed wave differential scheme [8, 9]

9.4.1 Operating principles

The basic principle of the wave differential scheme is that the direction to the fault can be determined by the sequence in which $S_{1R}^{(k)}$, $S_{2R}^{(k)}$, $k = 0, 1, 2$ exceed given setting levels (see Section 8.3.2), where $S_{1R}^{(k)}$, $S_{2R}^{(k)}$ are defined by eqns. 8.48.

For convenience they are rewritten below for each mode:

$$\left.\begin{aligned}
S_{1R}^{(0)} &= \Delta v_R^{(0)}(t) - Z_0^{(0)}\Delta i_R^{(0)}(t) \\
S_{2R}^{(0)} &= \Delta v_R^{(0)}(t) + Z_0^{(0)}\Delta i_R^{(0)}(t) \\
S_{1R}^{(1)} &= \Delta v_R^{(1)}(t) - Z_0^{(1)}\Delta i_R^{(1)}(t) \\
S_{2R}^{(1)} &= \Delta v_R^{(1)}(t) + Z_0^{(1)}\Delta i_R^{(1)}(t) \\
S_{1R}^{(2)} &= \Delta v_R^{(2)}(t) - Z_0^{(2)}\Delta i_R^{(2)}(t) \\
S_{2R}^{(2)} &= \Delta v_R^{(2)}(t) + Z_0^{(2)}\Delta i_R^{(2)}(t)
\end{aligned}\right\} \qquad (9.9)$$

9.4.2 Basic description of the scheme

In common with the RALDA system, the earliest work was done using a largely analogue implementation, and Figure 9.4 shows a block schematic diagram of an analogue prototype scheme. The currents and voltages derived from primary transformers are taken to their respective mixing circuits. In these circuits the currents and voltages are decomposed into their respective modal time variations according to eqns. 9.6. The modal currents are passed through surge replica circuits, where each modal component is multiplied by its corresponding replica surge impedance, which in turn results in a signal waveform $R_0^{(k)}i_R^{(k)}(t)$, $k = 0$, 1, 2. The modal voltage variations $v_R^{(k)}$ and those formed from the modal current components $(R_0^{(k)}i_R^{(k)}(t))$ are then mixed in the signal-mixing circuit/ analogue interface block to form the composite signals $v_R^{(k)} - R_0^{(k)}i_R^{(k)}(t)$, $k = 0$, 1, 2. It is important to note that at this point in the process the composite signals contain both steady-state and superimposed quantities.

Each composite signal is passed through a superimposed component extraction circuit, where the superimposed components are extracted from the total time variation of the composite signals; Figure 9.5(a) shows an analogue technique that was used to achieve this. Fig 9.5(b) illustrates the steady-state and the superimposed waveforms associated with an arbitrary chosen total waveform, from which it is evident that the superimposed component can be obtained over a limited period of time by delaying the total waveform by an integer multiple of the period $T = 1/f_0$. The use of a time delay of one cycle of nominal system is generally satisfactory, and Figure 9.5(c) shows the circuit waveforms that are often obtained. In particular it shows how the circuit of Figure 9.5(a) produces the superimposed component for one power frequency period (T) after the latter becomes finite.

The composite signals due to the superimposed quantities are then taken to the sequence detector block, which examines the direction of the fault in accordance with the sequence in which the discriminant signals of eqn. 9.9 exceed a threshold ($\pm V_s$), as explained in Chapter 8. Figure 9.6 illustrates in more detail the basis of the operation of the sequence-detector circuits for the mode 1 ($k = 1$) channel following forward and reverse faults.

Each level detector in the circuit of Figure 9.6(a) is arranged to give an output when the magnitude of the input signal exceeds a preset pickup or setting

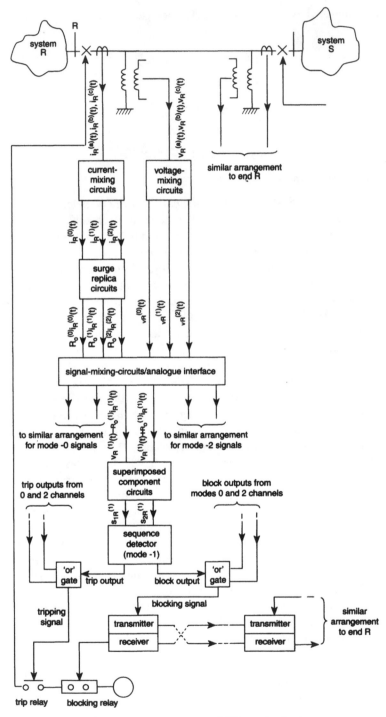

Figure 9.4   Block schematic diagram for wave-differential scheme

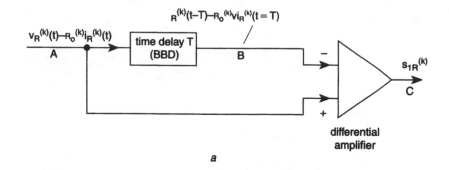

$$R^{(k)}(t-T)-R_0^{(k)}vi_R^{(k)}(t=T)$$

$$v_R^{(k)}(t)-R_0^{(k)}i_R^{(k)}(t)$$

A

time delay T
(BBD)

B

$-$

$+$

$s_{1R}^{(k)}$

C

differential
amplifier

a

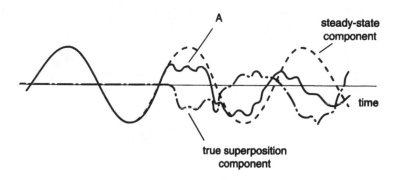

A

steady-state
component

time

true superposition
component

b

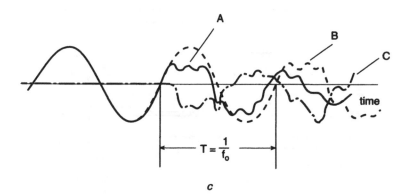

A

B

C

time

$$T = \frac{1}{f_0}$$

c

Figure 9.5 Basis of arrangement for extracting superimposed components

    (*a*) circuit arrangement
    (*b*) relationship of steady and superimposed components to total signal
       variation
    (*c*) waveform illustrating circuit action

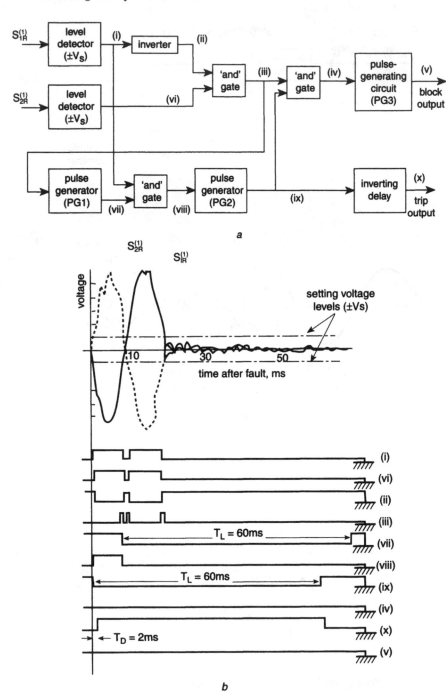

Figure 9.6   Sequence detector arrangement and function

    (*a*) schematic arrangement of mode 1 channel
    (*b*) circuit function for forward faults

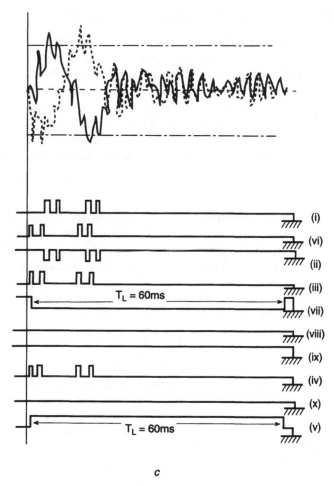

Figure 9.6   *Sequence detector arrangement and function (continued)*
(*c*) circuit function for reverse faults

voltage V_S. On the other hand, an input to the pulse-generating circuits PG1 or
PG2 causes their output to fall to zero for a preset time T_L. The latter time
determines how long the detector is latched following a forward or reverse fault,
thus ensuring that any subsequent change in the input signals is ignored.
Similar considerations apply to PG3, but this circuit is arranged so that its
output is normally zero and rises for a time of 60 ms following an input. The
tripping signal has to be delayed to enable the local blocking relay to operate on
receipt of a carrier blocking signal generated by the relay at the other end of the
line following an out-of-zone fault. This is achieved by the inverting delay
circuit which delays the trip signals by the required time.

9.4.3 Digital implementation of wave differential scheme [9]

The simple sequential threshold decision technique adopted in the analogue UHS wave differential scheme provides only a limited degree of security and dependability. Particular problems emerged largely from difficulties in guaranteeing satisfactory operation when the relaying signals were corrupted by noise or affected by primary devices in some very-high-noise substation environments. To improve performance and take advantage of developments in digital technology, the UHS wave differential scheme and commercial equipments derived from them have been engineered using digital signal processing techniques.

Operating principle
The operating principles adopted for the digitised UHS scheme are basically the same as those used in the analogue UHS scheme. For ease of explanation, consider the superimposed single-phase equivalent circuit shown in Figure 9.7. The composite superimposed signals used are those defined in eqns. 8.44, rewritten below. As also explained previously, the actual signals used employ modal components but for ease of explanation only the equivalent single phase form of signals repeated in eqn. 9.10 will be considered.

$$S_{1R}(t) = \Delta v_R(t) - R_0 \Delta i_R(t)$$
$$S_{2R}(t) = \Delta v_R(t) + R_0 \Delta i_R(t)$$

(9.10)

Signal magnitude comparison criteria were used to enable the directional decision to be based on several consecutive samples of the composite signals of eqn. 9.10, thereby improving relay security. This is desirable because, for a reverse fault behind the relay location in Figure 9.7, the simple signal criterion of eqn. 9.11 only holds for twice the wave transit time (l/c) between the line ends. In the case of a forward fault at a distance x_F from the relay location considered in Figure 9.7, the signal criterion of eqn. 9.12 holds only for so long

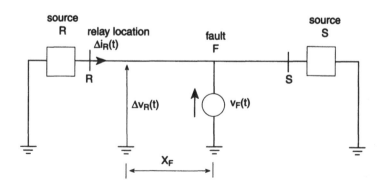

Figure 9.7   Simple superimposed component faulted circuit model

as reflections from within the source network R take to return to the relay location.

$$|S_{2R}(t)| > |S_{1R}(t)| \qquad (9.11)$$

$$|S_{1R}(t)| > |S_{2R}(t)| \qquad (9.12)$$

It will thus be apparent that in some circumstances the discriminative criteria eqns. 9.11 and 9.12 are valid only for relatively short periods after fault inception. However, they can be made to hold for extended periods of time by decreasing the influence of travelling wave components. This is illustrated in Figure 9.8, which shows the superimposed voltage and current due to a forward solid fault on the system shown in Figure 9.7 but with the line being represented by lumped elements, i.e. travelling waves are neglected. If the distributed nature of the line is ignored, as is often the case in practice, then the resulting travelling-wave components would simply be added to the superimposed power frequency and exponential transient components. Figure 9.8 shows that, when the travelling waves are absent, the forward directional criterion of eqn. 9.12 would hold for approximately one half cycle and one quarter cycle of the power frequency for faults at zero and maximum points on the wave, respectively. Specially designed digital filters are therefore used to decrease the influence of travelling-wave components after the initial post-fault period, thereby extending the time for which the discriminative criteria hold.

Variable thresholds, against which the filtered directional wave signals are compared, are used. These are adjusted in line with the prevailing noise in substation environments. This approach allows a sample-by-sample counting regime to be implemented, which also takes account of any momentary high-level noise bursts above the basic threshold, and therefore improves relay integrity. It also improves the response where significant signal components introduced by capacitor voltage transducer transients are generated and ensures that fast relay recovery is achieved.

9.4.4 General description of the digital relay

Figure 9.9 shows the block diagram of the digital form of the wave-differential relay. It can be seen that the design of the relay is based on using detectors for two modes of propagation only, thereby simplifying the hardware requirement. The primary transducer signals are converted, via the relay interface circuits, into analogue form for electronic pre-filtering and modal mixing before conversion to digital form for subsequent digital filtering and decision processing. After the modal mixing analogue stage, the two mode voltage and current signal pairs are processed as far as the directional decision stage in independent channels, and the resulting forward and reverse directional decisions are combined as shown in a common scheme logic processor. Adjustable gains within the analogue voltage and current channel facilitates the sensitivity

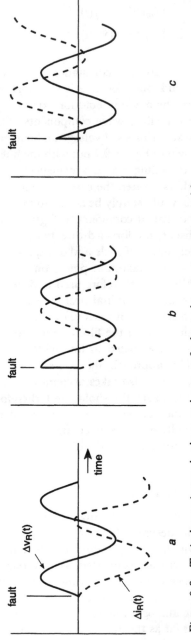

Figure 9.8   Typical superimposed voltage and current for lumped parameter system model

(a) fault at zero voltage
(b) fault at voltage maximum
(c) fault at 145° after voltage zero

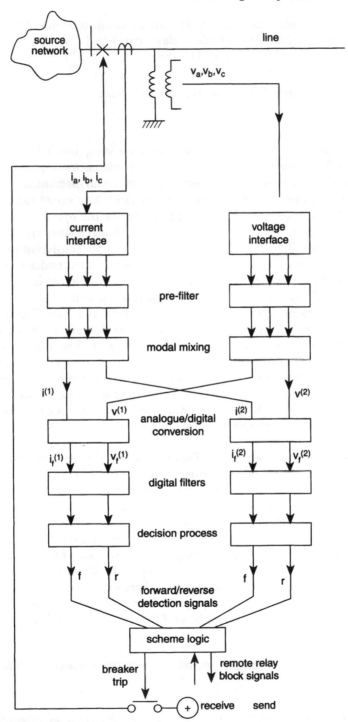

Figure 9.9   Digital relay block diagram

selection for particular applications. The arrangement of Figure 9.9 forms the basis of the digital algorithmic relay developed by GEC Alsthom Measurements, type designated LFDC.

9.5 Discriminant function based scheme

9.5.1 Operating principles [10, 11]

The operating principle of this scheme is based on using discriminant functions (see Section 8.5). This can best be explained with reference to Figure 9.10 where the table shown is constructed by calculating the modal discriminant functions using eqns. 8.71 and 8.72 for different types of fault. The values calculated are normalised with respect to V_{RMS}^2, i.e. the square of the operating voltage, to produce 0s and 1s. This table is based on the Karrenbauer transformation described by eqn. 8.28. It will be noted that this table consists of three main blocks marked phase 'a', 'b' and 'c'. The values tabulated under these blocks are obtained by assuming that the basis of the transformations is phase 'a', 'b' and 'c', respectively. The same procedure could be applied using the Wedepohl and Clark transformations described in eqns. 8.27 and 8.29. Figure 9.10(b) shows a flowchart that explains how the phase selection and fault classification can be achieved using modal discriminant functions. This can be achieved using the components $D_{B|a}^{(1)}$, $D_{B|a}^{(2)}$, $D_{F}^{(0)}$, $D_{F|b}^{(1)}$, $D_{F|a}^{(1)}$, $D_{F|a}^{(2)}$, where $D^{(k)}$ is the k modal discriminant function, and the first subscript is used to indicate whether it is a forward (F) or backward (B) function. The second subscript is used to refer to the phase on which the basis of the transformation is assumed. If the second subscript is omitted, the discriminant function is due to a three-phase fault condition.

As shown in the flowchart of Figure 9.10(b), the forward discriminant components $D_{F|a}^{(1)}$ and $D_{F|b}^{(1)}$ are first compared with a threshold value. A forward fault is indicated if either of the two components exceeds the threshold. Under this condition, the phase selection and fault classification are determined from the four forward components listed above and in accordance with the logic shown on the right-hand side of the flowchart. However, if none of the forward discriminant components $D_{F|a}^{(1)}$ or $D_{F|b}^{(1)}$ exceeds the specified threshold, the backward components $D_{B|a}^{(1)}$ and $D_{B|a}^{(2)}$ are checked. If either of them exceeds the threshold a backward fault is indicated, otherwise the system is considered sound. In the latter case the above procedure is repeated for the next set of samples. The measurements must be prefiltered, to avoid undue sensitivity to low levels of noise.

9.6 Superimposed component trajectory based scheme [12]

9.6.1 Basic principles

The direction of fault can be determined from the sense of rotation of the trajectories resulting from the superimposed quantities, in such a way that

fault →		L–G			L–L			L–L–G			3LS
basis	D's	a	b	c	a–b	b–c	c–a	a–b	b–c	c–a	
Ph "a"	D$^{(0)}$	1	1	1	0	0	0	1	1	1	0
	D$^{(1)}$	1	1	0	1	1	1	1	1	1	1
	D$^{(2)}$	1	0	1	1	1	1	1	1	1	1
Ph "b"	D$^{(0)}$	1	1	1	0	0	0	1	1	1	0
	D$^{(1)}$	0	1	1	1	1	1	1	1	1	1
	D$^{(2)}$	1	1	0	1	1	1	1	1	1	1
Ph "c"	D$^{(0)}$	1	1	1	0	0	0	1	1	1	0
	D$^{(1)}$	1	0	1	1	1	1	1	1	1	1
	D$^{(2)}$	0	1	1	1	1	1	1	1	1	1

a

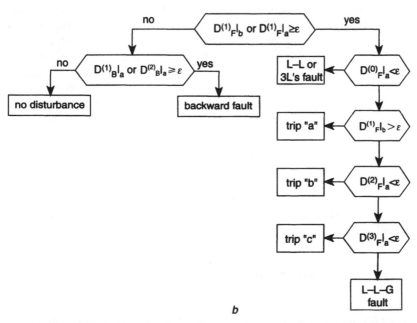

b

Figure 9.10   Fault detection, directional discrimination, phase selection, and fault classification based on the Karrenbauer transform [11]

(*a*) truth table
(*b*) process flow chart

forward faults give rise to a clockwise direction of rotation, while reverse faults give rise to trajectories with anticlockwise rotation.

9.6.2 Sense of trajectories versus fault direction

It has been shown in Section 8.3.4 that the behaviour of the superimposed voltage and current at the relay point can be described by an elliptical trajectory in the Δv_R versus $R\Delta i_R$ plane, where R is an arbitrary scale factor. The trajectory itself is mathematically represented by eqn. 8.55, rewritten below. (Note that subscript R is omitted in the following analysis.)

$$\frac{\Delta v^2(t)}{(\Delta V)^2} + \frac{[R\Delta i(t) - R\Delta I \cos \theta]^2}{[R\Delta I]^2} = 1 \tag{9.13}$$

Figure 9.11 shows the elliptical trajectories due to forward and reverse faults with respect to a relay located at the R end of the line R–S. Figure 9.11(a) illustrates the superimposed component trajectories due to forward faults. These trajectories possess a clockwise sense of rotation and they depart from the Δv axis, either into quadrant II or IV, depending on whether the fault occurs on the positive or negative half cycle of the prefault voltage waveform.

On the other hand, Figure 9.11(b) illustrates the trajectories due to a reverse-fault condition. Here it can be seen that the trajectories rotate in an anticlockwise direction and initially enter quadrant I or III, again depending on which half cycle of the voltage waveform the fault occurs in. Therefore, a basic technique to determine the direction of the fault using the above properties of the superimposed component trajectories is to introduce threshold boundaries in the Δv–$R\Delta i$ plane. Once the trajectory has crossed the boundary, the relay may reset or adjust itself to the postfault sinusoidal voltages and currents; a typical boundary is shown in Figure 9.12. A simplified method of boundary checking involves taking linear combinations of $\Delta v(t)$ and $R\Delta i(t)$ and checking against fixed values. For example, if the linear combination of $R\Delta i(t)$–$\Delta v(t)$ first reaches some threshold value, say $+E_0$ or $-E_0$, a forward fault will be indicated. Mathematically this is expressed as

$$R\Delta i(t) - \Delta v(t) = \pm E_0 \tag{9.14}$$

However, if the linear combination of $R\Delta i(t) + \Delta v(t)$ first reaches the threshold values, then a reverse fault will be indicated. More explicitly this can be expressed as

$$R\Delta i(t) + \Delta v(t) = \pm E_0 \tag{9.15}$$

Boundaries such as hyperbolas, circles, ellipses or piecewise linear curves can also be used.

9.6.3 Extension of trajectories approach to signals including travelling-wave components

The superimposed component trajectories which have been discussed so far are confined to quasi-steady-state conditions. The same principles can, however, be

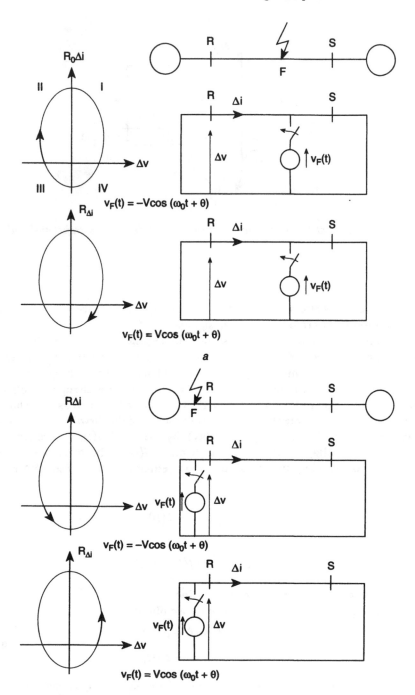

Figure 9.11   Elliptical trajectories due to

(a) forward faults
(b) reverse faults

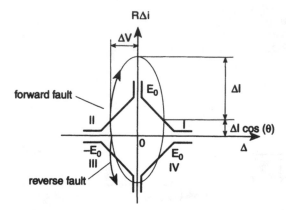

*Figure 9.12   Typical boundary used to determine fault direction in trajectory-based relay
[12]*

extended to transient conditions and particularly to the case when travelling-wave components are present in the signals.

It will be recalled from the analysis of a single-phase transmission line having constant distributed inductances and capacitances (see Section 8.2.1) that the voltage and current is made up of forward and backward travelling waves. Now let us consider the superimposed signals resulting from a change in voltage v_F due to a fault at point F on a distributed single-phase line such as that shown in Figure 9.13. If we denote the instantaneous values of the forward and backward travelling waves (as seen by the relay) by $f(t)$ and $b(t)$, respectively, the superimposed voltage and current $\Delta v(t)$ and $\Delta i(t)$ at the relay point can be expressed by solving the resultant wave equation using the D'Alembert approach:

$$\Delta v(t) = \frac{1}{2} \left(f(t) - b(t) \right)$$

$$\Delta i(t) = \frac{1}{2Z_0} \left(f(t) - b(t) \right)$$

(9.16)

where Z_0 is the surge impedance of the line.

By solving eqns. 9.16 for $f(t)$ and $b(t)$, we obtain

$$f(t) = Z_0 \Delta i(t) + \Delta v(t)$$

$$b(t) = Z_0 \Delta i(t) - \Delta v(t)$$

(9.17)

A logical approach to assess the location of the fault with respect to a relay location is to detect which of the two waves first reaches a given threshold constant E_0. Therefore if the wave signal $b(t)$ associated with the backward travelling wave reaches the threshold value $\pm E_0$, then a fault in the forward

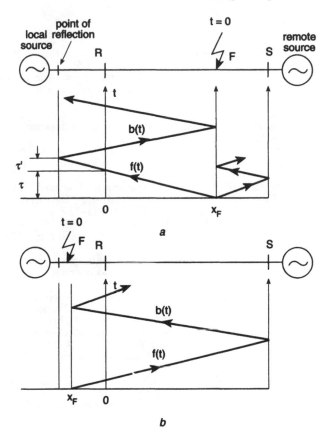

Figure 9.13   Travelling waves generated by a fault

(a) internal fault
(b) external fault

Zero fault resistance assumed [12]

direction is assumed. This condition can be expressed mathematically as

$$Z_0\Delta i(t) - \Delta v(t) = \pm E_0 \qquad (9.18)$$

On the other hand, if the wave signal $f(t)$ associated with the forward travelling wave increases beyond the absolute value of E_0, a reverse fault will be indicated. This condition is given by the following equation:

$$Z_0\Delta i(t) + \Delta v(t) = \pm E_0 \qquad (9.19)$$

The operating conditions described by eqns. 9.18 and 9.19 are identical to eqns. 9.14 and 9.15, which describe the slope boundaries previously discussed for the assumed quasi-steady-state condition provided that the parameter Z_0 is chosen as $Z_0 = R$. This is achieved in a practical implementation by using a current mimic impedance R_0 in deriving the signal component due to the superimposed current $\Delta i(t)$.

9.7 References

1 TAKAGI, T. *et al.*: 'Fault protection based on travelling wave theory, Part I: theory', PES Summer Meeting, Mexico, 1977, Paper A77, 750–753

2 TAKAGI, T. *et al.*: 'Fault protection based on travelling wave theory, Part II: sensitivity analysis and laboratory test', PES Winter Meeting, 1978, Paper A78, 220–226

3 AGGARWAL, R.K., and JOHNS, A.T.: 'A differential line protection scheme for power systems using composite voltage and current measurements', *IEEE Trans.*, 1989, **PWRD-4**, pp. 1595–1601

4 AGGARWAL, R.K., and JOHNS, A.T.: 'A new differential protection scheme for power systems using composite voltage and current measurements', 23rd UPEC, 1988, Paper A3.4

5 AGGARWAL, R.K., and JOHNS, A.T.: 'New approach to teed feeder protection using composite current and voltage signal comparison', Developments in power system protection (IEE Conf. Publ. 302, 1989), pp. 125–129

6 CHAMIA, M., and LIBERMAN, S.: 'Ultra-high speed relay for EHV/UHV transmission lines—development, design and application', *IEEE Trans.* 1978, **PAS-97**, pp. 2104–2116

7 YEE, M.T., and ESZTERGALYOS, J.: 'Ultra high speed relay for EHV/UHV transmission lines—installation-staged fault tests and operational experience', *ibid*, pp. 1814–1825

8 JOHNS, A.T.: 'New ultra-high-speed directional comparison technique for the protection of EHV transmission lines', *IEE Proc. C* 1980, **127**, pp. 228–239

9 JOHNS, A.T., MARTIN, M.A., BARKER, A., WALKER, E.P., and CROSSLEY, P.A.: 'A new approach to EHV directional comparison protection using digital signal processing techniques', *IEEE Trans.* 1986, **PWRD-1**, pp. 24–34

10 MANSOUR, M.M., and SWIFT, G.W.: 'A multi-microprocessor based travelling wave relay—theory and realization', *ibid*, pp. 273–279

11 MANSOUR, M.M., and SWIFT, G.W.: 'Design and testing of a multi-microprocessor travelling-wave relay', *ibid*, pp. 74–82

12 VITINS, M.: 'A fundamental concept for high speed relaying', *IEEE Trans.* 1981, **PAS-100**, pp. 163–173

Chapter 10

Digital differential protection of transformers

10.1 Introduction

This Chapter gives a brief general review of the principles of transformer differential protection. This is followed by an explanation of the application of digital techniques and the algorithms that have been developed specifically for application to transformer protection. The algorithms covered include finite-duration impulse (FIR) filters, least-squares curve fitting, the digital Fourier algorithm and the flux-restrained current differential algorithm.

Finally, the basic hardware arrangement for implementing digital techniques to the protection of transformers is described. It is, however, important to note that closely similar techniques can be applied to the protection of generators, although, in this case, the transformation ratio of currents is the same on each side of the protected zone.

10.2 Principles of transformer protection [1–4]

10.2.1 Basic principles

The most commonly encountered transformer protection arrangement is based on the differential current principle. This can be illustrated by reference to Figure 10.1, in which the primary and secondary currents (I_p and I_s) are compared after being reduced by their corresponding current transformers. The

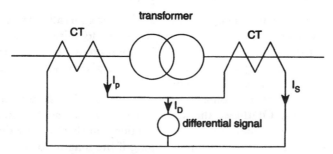

Figure 10.1   Basic principle of transformer differential protection

primary and secondary current transformers are connected such that ideally, under healthy conditions, only circulating currents flow and the differential signal I_D is zero. If an internal fault occurs within the transformer, the balance between the primary and secondary currents is disturbed and a differential current signal $(I_D = I_p - I_s)$ then causes the relay to operate.

10.2.2 Biased differential relaying

If the transformer is equipped with a tap changer, the imbalance between the primary and secondary currents introduced by the variation of the turns ratio can be great enough to cause malfunction during external fault conditions. Imbalance under healthy conditions can also be caused by mismatch between the current transformers or by saturation. It is therefore important that this situation should be taken into account in the engineering of transformer differential protection systems.

A common way of overcoming such problems is to bias the operation of the relay by deriving a biasing signal equal to the sum of the currents measured on each side of the transformer. This is illustrated in Figure 10.2(*a*) together with a typical operating characteristic (Figure 10.2(*b*)). It is evident from Figure 10.2(*b*) that the differential current required to operate the relay increases with the bias current. Thus the stability of the relay under external fault conditions is ensured. For example, if a heavy through fault (healthy) condition exists, the differential current may take a finite value that is nevertheless much smaller than the associated bias current. The restraining characteristic of Figure 10.2(*b*) is arranged to allow a significant level of differential current to occur without causing tripping under healthy conditions. Conversely, a faulty condition causes the bias current to be relatively small, on account of the reversal of the current measurand I_s, thus ensuring that tripping occurs.

In some cases the operating current is expressed as a percentage of the restraining current, and such relays are commonly known as percentage-biased differential relays. Many practical designs try to derive differential and bias signals proportional to the fundamental or power frequency components of measured current, which is why in Figure 10.2(*b*) the measurands are subscripted.

10.2.3 Harmonic-restrained differential relay [5, 6]

The differential protective scheme described above is generally very satisfactory under normal operating conditions, but is prone to false operation during energisation of the transformer. Under this condition, an inrush current flows only through one winding (the winding being energised), which consequently appears to the relay as an internal fault.

This problem can be overcome by using the fact that inrush currents usually contain harmonics. Of these, the second harmonic is usually predominant under all energisation conditions. The protection can therefore be designed so that it is prevented (restrained) from tripping if the magnitude of the second-harmonic component of operating current is greater than a certain prescribed percentage of the fundamental component. Fifth harmonics, which are generated under over-excitation conditions (caused, for example, by temporary overvoltages) are also used in some equipment to prevent the relay from tripping under energisation conditions.

It will be apparent that harmonic-restrained current differential transformer protection basically uses signals formed from fundamental (or power frequency) and harmonic components. The power-frequency components predominate when an internal fault occurs, and the harmonic and bias signals are used to restrain the protection during through faults, switching operations and during over-excitation conditions.

With reference to Figure 10.2(*a*), the differential current I_D is commonly formed from the fundamental frequency components of the primary and secondary currents I_{1p} and I_{1s}, respectively, such that

$$I_D = I_{1p} - I_{1s} \qquad (10.1)$$

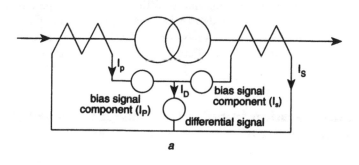

a

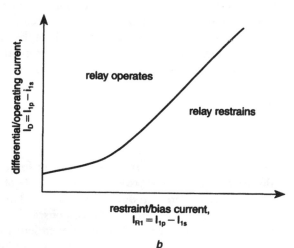

b

Figure 10.2   Biased differential relay

    (*a*) basic arrangement
    (*b*) typical operating characteristic

The restraining signal (I_{R1}), on the other hand, is commonly formed from the fundamental components of the primary and secondary components such that

$$I_{R1} = I_{1p} + I_{1s} \tag{10.2}$$

The other restraint signals are the second-harmonic restraint I_{R2} and fifth-harmonic restraint current I_{R5}, which are commonly formed from the magnitudes of differential currents of any second and fifth harmonics detected. A commonly used technique involves summing any such harmonic components in each phase:

$$I_{R2} = |I_{D2a}| + |I_{D2b}| + |I_{D2c}| \tag{10.3}$$

and

$$I_{R5} = |I_{D5a}| + |I_{D5b}| + |I_{D5c}| \tag{10.4}$$

where I_{D2a}, I_{D2b} and I_{D2c} are the differential currents due to second-harmonic components in phases 'a', 'b' and 'c', respectively, and I_{D5a}, I_{D5b} and I_{D5c} are the corresponding differential currents due to fifth-harmonic components.

Tripping and restraining signals can be derived using analogue techniques (as in the case of electromagnetic relays or solid-state relays) but this Chapter deals with the derivation and use of such signals using digital techniques.

10.3 Finite-duration impulse response filter based algorithms [7, 8]

These algorithms are based on using finite-duration impulse response (FIR) filters to estimate the magnitude of the fundamental and second-harmonic components. The estimates are then used to calculate the ratio of the magnitude of second harmonic to fundamental. If this ratio is found to be greater than a certain value, an inrush current condition is assumed. This algorithm is often designed to avoid multiplication and division as these operations are rather unwieldy to perform at both software and hardware levels.

10.3.1 FIR filter characteristics

The filters used are assumed to be characterised by means of finite-duration impulse responses of duration equal to one power-frequency cycle of period T and having values of $+1$ or -1 at any instant during that period. The choice of ± 1 for the impulse response avoids the need to find the responses of the filters to arbitrary inputs and eliminates the time-consuming multiplications that would otherwise be required, although in many modern DSP processors this is not necessary.

To estimate the magnitude of fundamental and second-harmonic components of the input, four filters are required: two for the fundamental and two for the second-harmonic component. Their impulse responses are $S_1(t)$, $C_1(t)$, $S_2(t)$ and $C_2(t)$ (Figure 10.3), which in turn are defined by the following equations:

$$S_1(t) = \begin{cases} +1 & 0 \leqslant t \leqslant T/2 \\ -1 & T/2 \leqslant t \leqslant T \end{cases} \tag{10.5}$$

$$C_1(t) = \begin{cases} +1 & 0 \leqslant t \leqslant T/4, \ 3T/4 \leqslant t \leqslant T \\ -1 & T/4 \leqslant t \leqslant 3T/4 \end{cases} \tag{10.6}$$

$$S_2(t) = \begin{cases} +1 & 0 \leqslant t \leqslant T/4, \ T/2 \leqslant t \leqslant 3T/4 \\ -1 & T/4 \leqslant t \leqslant T/2, \ 3T/4 \leqslant t \leqslant T \end{cases} \tag{10.7}$$

$$C_2(t) = \begin{cases} +1 & 0 \leqslant t \leqslant T/8, \ 3T/8 \leqslant t \leqslant 5T/8, \ 7T/8 \leqslant t \leqslant T \\ -1 & T/8 \leqslant t \leqslant 3T/8, \ 5T/8 \leqslant t \leqslant 7T/8 \end{cases} \tag{10.8}$$

It will be apparent that $S_1(t)$, $C_1(t)$ are impulse responses for the sine and cosine parts of the fundamental component, and $S_2(t)$, $C_2(t)$ are impulse responses for the sine and cosine parts of second-harmonic components. Eqn. 2.73 can be applied to eqns. 10.5–10.8 to find the system frequency responses of the four filters, which are consequently given by

$$F_{S1}(\omega) = \frac{2}{j\omega} e^{-j\omega T/2} \left(\cos\left(\frac{\omega T}{2}\right) - 1 \right) \tag{10.9}$$

$$F_{C1}(\omega) = \frac{2}{j\omega} e^{-j\omega T/2} \left(\cos\left(\frac{\omega T}{2}\right) - 2 \sin\left(\frac{\omega T}{4}\right) \right) \tag{10.10}$$

$$F_{S2}(\omega) = \frac{2}{j\omega} e^{-j\omega T/2} \left(\cos\left(\frac{\omega T}{2}\right) - 2 \cos\left(\frac{\omega T}{4}\right) + 1 \right) \tag{10.11}$$

$$F_{C2}(\omega) = \frac{2}{j\omega} e^{-j\omega T/2} \left(\sin\left(\frac{\omega T}{2}\right) - 2 \sin\left(\frac{3\omega T}{8}\right) + 2 \sin\left(\frac{\omega T}{8}\right) \right) \tag{10.12}$$

where $T = 2\pi/\omega_0$ is one period of the system frequency f_0.

The amplitudes of the above frequency responses plotted against the ratio ω/ω_0 are shown in Figure 10.4.

10.3.2 Extraction of fundamental and second-harmonic components

To extract the fundamental component of a current, it is necessary to determine the responses of the filters to the transformer current. The responses of the filters 1, 2, 3 and 4 to transformer current after one cycle of power frequency ($t = T$) can be found by convolving the impulse responses of these filters with the transformer current (as an input to the filters) and evaluating at time $t = T$.

As always in digital protection, the current inputs to the filters are in the form of samples, and since the impulse responses of the FIR filters have the values of either $+1$ or -1 at any instant of time (as defined by eqns. 10.5–10.8), the multiplication operation required by the convolution process is simply a sign

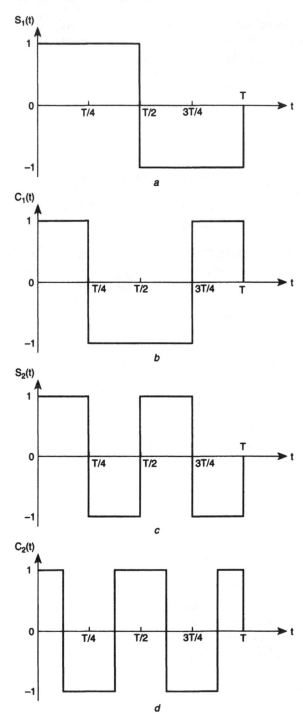

Figure 10.3   Impulse response of fundamental and second harmonic FIR filters

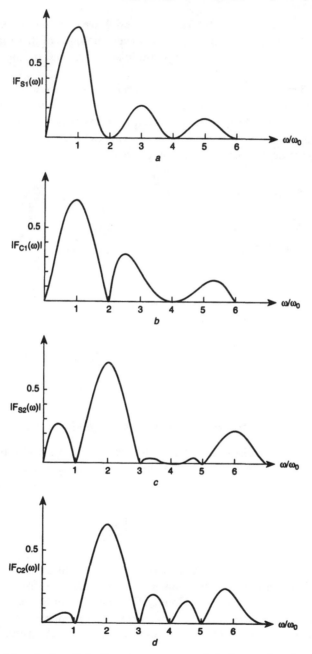

Figure 10.4 *Magnitudes of the frequency responses of fundamental and second-harmonic filters*

    (*a*) filter 1
    (*b*) filter 2
    (*c*) filter 3
    (*d*) filter 4

change. Thus the outputs of filters 1, 2, 3 and 4 are obtained by summing the samples of their inputs over one cycle. However, the signs of these samples must be modified in accordance with the impulse responses of the filter under consideration.

Assume N is the number of samples per cycle of the current $i(t)$ and is chosen as a multiple of eight. In this case, the time between successive samples is $\Delta t = 2\pi/(N\omega_0)$ and $i_k = i(t_k)$ is the kth sample at any time $t = k\Delta t$. The result of time-discrete convolution of the samples i_k with the impulse responses defined by the previous equations will then be given as

$$S_1(t) = \sum_{k=1}^{N/2} [i_k - i_{k+N/2}] \tag{10.13}$$

$$C_1(t) = \sum_{k=1}^{N/4} [i_k - (i_{k+N/4} + i_{k+N/2}) + i_{k+3N/4}] \tag{10.14}$$

$$S_2(t) = \sum_{k=1}^{N/4} [i_k - i_{k+N/4} + i_{k+N/2} - i_{k+3N/4}] \tag{10.15}$$

$$C_2(t) = \sum_{k=1}^{N/8} [i_k - (i_{k+N/8} + i_{k+N/4}) + (i_{k+3N/8} + i_{k+N/2})$$

$$- (i_{k+5N/8} + i_{k+3N/4}) + i_{k+7N/8}] \tag{10.16}$$

10.3.3 Discrimination between inrush and internal fault currents

The criteria used to distinguish between inrush currents and currents due to internal faults is based on evaluating the ratio ε, which is calculated from the ratio of the larger of the two components of each pair of filter outputs as described by eqns. 10.13–10.16 such that

$$\varepsilon = \frac{\max(|S_2|, |C_2|)}{\max(|S_1|, |C_1|)} \tag{10.17}$$

An internal fault, as distinct from a current inrush condition, is characterised by a high value of fundamental component of differential current and a relatively low value of second harmonic. Conversely, inrush conditions are associated with relatively high second harmonic current components. It has been suggested [7] that the values of ε for the detection of internal faults are

$$0 \leqslant \varepsilon \leqslant 0.146 \qquad \text{for } X/R = 5$$

$$0 \leqslant \varepsilon \leqslant 0.093 \qquad X/R = 10$$

$$0 \leqslant \varepsilon \leqslant 0.054 \qquad X/R = 20$$

where X/R is the system reactance–resistance ratio.

A typical threshold value ε_0, which is convenient from a computational point of view, is $\varepsilon_0 = 0.125$. In such a case, any value ε which is greater than 0.125 would be taken to indicate an inrush condition and would inhibit operation of the protection.

10.4 Least-squares curve fitting based algorithms [9, 10]

Least-squares (LSQ) curve fitting methods have been discussed in detail in Chapter 6. These methods are basically used to extract the fundamental and/or any other harmonic components. Since the relative values of second-harmonic and fundamental components of the differential current play an important role in the transformer differential protection, LSQ techniques can be used either to detect the second harmonic in an inrush current (which is later used to block the differential protection) or to find the ratio between the fundamental and second-harmonic components of the differential current, which in turn is used to differentiate between inrush current and internal faults.

10.4.1 Basic assumptions and algorithm derivation

Since inrush currents can in fact contain significant harmonics up to the fifth, the differential current can be described by a waveform containing a decaying DC component, fundamental and harmonic components up to the fifth order:

$$i(t) = I_0 \, e^{-t/\tau} + \sum_{m=1}^{5} I_m \cos(m\omega_0 t + \theta_m) \qquad (10.18)$$

where τ is the time constant of any decaying DC component.

If the decaying DC component is approximated by the first two terms of its Taylor expansion, then the current sample $i(t_k)$ measured at t_k can be expressed as

$$i(t_k) = I_0 - (I_0/\tau)t_k + \sum_{m=1}^{5} (I_m \cos \theta_m) \cos m\omega_0 t_k + \sum_{m=1}^{5} (I_m \sin \theta_m) \sin m\omega_0 t_k$$

$$(10.19)$$

or

$$i_k = a_1 x_1 + a_2 x_2 + \cdots + a_{11} x_{11} + a_{12} x_{12}$$

with

$$a_1 = 1 \qquad\qquad\qquad\qquad x_1 = I_0$$

$$a_2 = t_k \qquad\qquad\qquad\qquad x_2 = -I_0/\tau$$

$$\left.\begin{array}{l} a_{n+2} = \cos n\omega_0 t_k \\ a_{n+7} = \sin n\omega_0 t_k \end{array}\right\} n = 1, \ldots, 5 \qquad \left.\begin{array}{l} x_{n+2} = I_n \cos \theta_n \\ x_{n+7} = I_n \sin \theta_n \end{array}\right\} n = 1, \ldots, 5$$

To solve for the unknowns $x_n(n = 1, \ldots, 12)$, m equations can be constructed from N current samples. The resulting equations can be written in matrix form:

$$\underbrace{A}_{N \times 12} \ \underbrace{X}_{12 \times 1} = \underbrace{i}_{N \times 1} \qquad (10.20)$$

or

$$X = Bi \qquad (10.21)$$

where $B = \{A^T \cdot A\}^{-1} \cdot A^T$ is the pseudoinverse of A, and A^T is the transpose of matrix A (see Section 2.5.1).

It follows that the real and imaginary parts of the fundamental and second harmonic can be calculated from eqn. 10.21:

$$I_1 \cos \theta_1 = x_3 = \sum_{n=1}^{N} b(3, n)i_n$$

$$I_1 \sin \theta_1 = x_8 = \sum_{n=1}^{N} b(8, n)i_n$$

$$\qquad (10.22)$$

$$I_2 \cos \theta_2 = x_4 = \sum_{n=1}^{N} b(4, n)i_n$$

$$I_2 \sin \theta_2 = x_9 = \sum_{n=1}^{N} b(9, n)i_n$$

where x_k is the kth element of vector X and $b(k, n)$ is the element of the kth row and nth column of matrix B.

Using the foregoing equations, the amplitude of the fundamental and the second harmonic can be calculated as

$$I_n = \sqrt{(I_n \cos \theta_n)^2 + (I_n \sin \theta_n)^2} \quad n = 1, 2 \qquad (10.23)$$

Similar techniques can be applied in extracting, where necessary, the fifth-harmonic components.

10.4.2 Basis of discrimination between inrush and internal fault currents

Discrimination between an inrush and an internal fault condition is often based on an inrush-detection algorithm. This algorithm is based on comparing the second-harmonic and fundamental components I_2 and I_1, which are present in the differential current. The values of these currents, determined according to

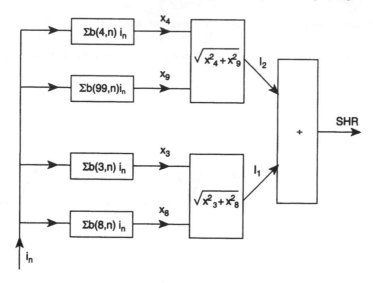

Figure 10.5   Schematic representation of the determination of the second harmonic (SH) ratio

i_n is the nth sample, $b(k, n)$ is the element of kth row and nth column of matrix **B** (see eqn. 10.21) and Σ is the sum from $n = 1$ to N, i.e. the number of samples per fundamental period

eqn. 10.23, are then used to calculate what is called the 'second harmonic ratio', which is defined by

$$\text{SHR} = \frac{I_2}{I_1} \qquad (10.24)$$

If the SHR is greater than a set value, an inrush current condition is assumed and tripping is prevented; otherwise an internal fault condition is assumed and a tripping signal is issued to isolate the transformer.

Figure 10.5 shows a diagrammatic representation of the four digital filters used to derive the real parts, i.e. x_3 and x_4, and imaginary parts (x_8 and x_9) of the fundamental and second-harmonic components used in calculating the SHR. The frequency characteristics of the four filters are shown in Figure 10.6 where it can be seen that the unwanted harmonics are filtered out.

10.5 Fourier-based algorithm [11]

10.5.1 Filtering of harmonics

This algorithm is based on the fact that the fundamental, second- and fifth-harmonic components contained within an inrush current can be digitally extracted using the Fourier approach given in detail in Section 5.2.1.

Now assume (as previously) that the current waveform is sampled N times per period of the fundamental, and let the samples be denoted by $i_k = i(k\Delta t)$. The real and imaginary parts of the nth harmonic (a_n and b_n) can be found by

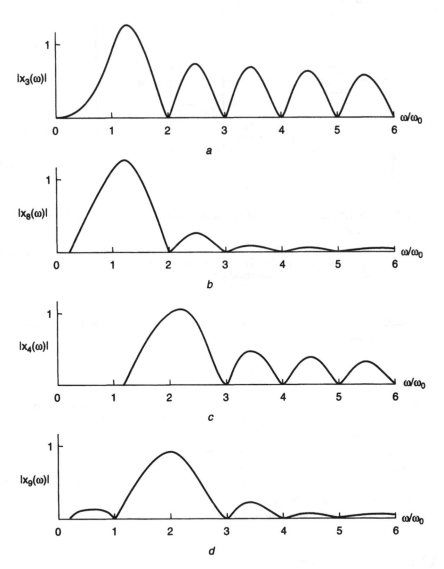

Figure 10.6   *The magnitudes of the frequency responses of digital harmonic extraction filters*

  (a) $x_3(\omega)$
  (b) $x_8(\omega)$
  (c) $x_4(\omega)$
  (d) $x_9(\omega)$

using eqns. 5.5 and 5.7. In terms of current samples starting at the rth sample, a_n and b_n can then be expressed as

$$a_n^{(r)} = \frac{2}{N} \sum_{k=r}^{r+N-1} i_k \cos n\left(\frac{2\pi k}{N}\right) \tag{10.25}$$

$$b_n^{(r)} = \frac{2}{N} \sum_{j=r}^{j+N-1} i_k \sin n\left(\frac{2\pi k}{N}\right) \tag{10.26}$$

The magnitude $|I_n^{(r)}|$ of the nth harmonic and its phase angle $\theta_n^{(r)}$ can thus be calculated as explained previously. Thus

$$|I_n^{(r)}| = \sqrt{(a_n^{(r)})^2 + (b_n^{(r)})^2}$$

$$\theta_n^{(r)} = \tan^{-1} \frac{b_n^{(r)}}{a_n^{(r)}} \tag{10.27}$$

For the transformer protection application, n takes the values of 1, 2 and 5 for fundamental, second and fifth harmonics, respectively. The result can be updated iteratively as each new sample becomes available. This is done by dropping the earliest sample and adding the new sample:

$$a_n^{(r+1)} = a_n^{(r)} + \frac{2}{N} [i_{N+r} - i_r] \cos n\left(\frac{2\pi r}{N}\right) \tag{10.28}$$

$$b_n^{(r+1)} = b_n^{(r)} + \frac{2}{N} [i_{N+r} - i_r] \sin n\left(\frac{2\pi r}{N}\right) \tag{10.29}$$

where i_r and i_{N+r} are the oldest and newest samples, respectively.

Having determined the fundamental, second- and fifth-harmonic components, the transformer protection is then implemented following the principles outlined in Section 10.2.3.

Other techniques have been reported in the literature to extract fundamental and other harmonic components [11]. These techniques give closely similar performance and include the use of rectangular transforms [12], Walsh functions [13] and the Haar function [14].

10.6 Flux-restrained current differential relay [15, 16]

This algorithm basically uses the flux-current relation of the transformer to obtain the restraint function. If the flux could be estimated correctly, it would provide a sound basis for detecting over excitation as well as magnetising inrush conditions. An important feature of this technique is that it requires fewer computations than that based on the Fourier analysis.

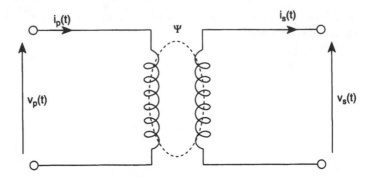

Figure 10.7   Two-winding single-phase transformer

10.6.1 Development of the algorithm

Consider a simple two-winding tranformer (basically the same analysis can be applied to a multi-winding transformer) such as that shown in Figure 10.7. It will be assumed that winding resistance is negligible. The relation between the primary applied voltage v_p, the primary current i_p and the mutual flux linkages ψ of the transformer is given by

$$L_p \frac{di_p(t)}{dt} + \frac{d\psi(t)}{dt} = v_p(t) \tag{10.30}$$

where L_p is the leakage inductance of the primary winding. By rearranging eqn. 10.30 and integrating from t_1 to t_2, we obtain the flux linkages at these times such that

$$\psi(t_2) - \psi(t_1) = \int_{t_1}^{t_2} v_p(t)dt - L_p[i_p(t_2) - i_p(t_1)] \tag{10.31}$$

Applying the trapezoidal rule to the integral part of eqn. 10.31 results in

$$\psi(t_2) \simeq \psi(t_1) + \frac{1}{2}(t_2 - t_1)[v_p(t_2) - v_p(t_1)] - L_p[i_p(t_2) - i_p(t_1)] \tag{10.32}$$

If the voltage and current waveforms are sampled such that the sampling interval is equal to Δt, then at the kth sample, eqn. 10.32 can be expressed using sample notation as

$$\psi_k = \psi_{k-1} + \frac{1}{2}\Delta t(v_{p,k} - v_{p,k-1}) - L_p(i_{p,k} - i_{p,k-1}) \tag{10.33}$$

where $i_{p,k}$, $v_{p,k}$ are the kth samples of the primary current and voltage.

Eqn. 10.33 can therefore be used to calculate the mutual linkage flux ψ of the transformer. On the other hand, the differential current can be calculated from the primary and secondary currents. Thus at time t_k, the differential current i_{dk} is given as:

$$i_{dk} = i_{p,k} - i_{s,k} \qquad (10.34)$$

where $i_{p,k}$ is as defined previously and $i_{s,k}$ is the kth sample of the secondary current. However, from transformer theory, the differential current i_{dk} is equal to the magnetising current of the transformer. Therefore, if it is assumed that the flux linkage computed by eqn. 10.33 is a true representation of the actual flux in the transformer, the samples of the differential current and flux linkage (i_{dk}, ψ_k) are expected to fall on the open-circuit magnetising curve of the transformer.

In the first phase of the algorithm, the principle of the percentage-biased differential relay is used to detect internal faults at every sampling interval. The second phase of the algorithm involves a simultaneous check on the location of the point (i_{dk}, ψ_k). If this point does not fall on the i_d–ψ curve of the transformer, a trip signal is issued.

The basis of this approach is explained in Figure 10.8(a), which shows an open-circuit magnetising characteristic of a transformer and the (i_{dk}, ψ_k) relationship for an internal fault inside the transformer. For the internal fault condition, the terminal voltage (and hence ψ_k) is much smaller than is the case under inrush conditions. Consequently there are distinct regions in the i–ψ plane that define the fault or non-fault status of the transformer.

The above technique works satisfactorily when the residual flux in the core of the transformer is close to zero. Practically this is not always the case, and consequently the i_d–ψ characteristic of the transformer may vary as shown in Figure 10.8(b). This situation makes it impossible to differentiate between fault and no-fault regions in the i–ψ plane if the flux linkages are computed according to eqn. 10.33. This is because the estimated value is subjected to an error equal to the value of the residual flux linkages.

One way of overcoming this problem is to use a restraining function which is determined by the slope $d\psi/di$ rather than the flux ψ itself. From eqn. 10.33 we then obtain

$$\left(\frac{d\psi}{di}\right)_k \approx \frac{\psi_k - \psi_{k-1}}{i_k - i_{k-1}} = \frac{1}{2}\,\Delta t\,\frac{V_{p,k} + V_{p,k-1}}{i_{p,k} - i_{p,k-1}} - L_p \qquad (10.35)$$

Figure 10.9 shows that there are two regions in the $d\psi/di$–i plane. The first region corresponds to a fault condition or an operation in the saturated part of the magnetisation curve. The second region, which is significantly removed from the first, designates an operation on the magnetising curve in the unsaturated part. It has been observed that for internal fault conditions the current samples and corresponding flux derivative $d\psi/di$ remain continuously in region 1. On the other hand, during inrush conditions, they alternate between the two regions. This phenomenon can be used to create an index of restraint k, which is increased each time a sample pair $(i_k, (d\psi/di)|_k)$ falls in region 1. However, the index is decreased whenever the sample pair enters region 2. The index is also constrained to remain always positive. Thus

$$k_r = k_r + 1 \quad \text{if the differential current indicates a}$$

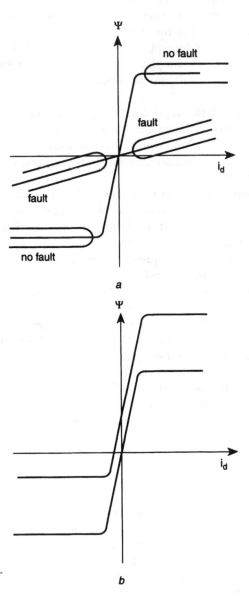

Figure 10.8   Transformer magnetising curve

    (*a*) fault and non-fault region
    (*b*) effect of remanent flux

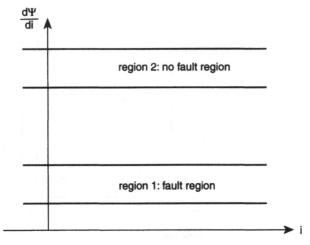

Figure 10.9 Fault and non-fault regions in the dψ/di–i plane

$$k_r = \begin{cases} k_r - 1 \\ \\ k_r \end{cases} \quad \begin{matrix} \text{trip condition and if the pair} \\ [i_k, \, (d\psi/di)|_k] \text{ is in region 1} \\ \\ \text{if } k_r > 0, \text{ and the pair } [i_k, \, (d\psi/di)|_k] \\ \text{is in region 2} \\ \text{if } k_r = 0 \end{matrix} \qquad (10.36)$$

The value of k_r is found to increase almost monotonically for fault conditions while it never reaches greater than a threshold value $k_{r\,\text{max}}$ for all non-fault conditions. However, the threshold $k_{r\,\text{max}}$ is found to depend, among other things, on the sampling rate. Therefore $k_{r\,\text{max}}$ must be determined experimentally.

10.7 Basic hardware of microprocessor-based transformer protection [17]

Figure 10.10(a) shows the connection of a microprocessor-based protective scheme to a three-phase, two winding Δ/Y transformer. The three-phase currents at the primary and secondary sides are first reduced using current transformers of suitable ratio, then connected to the microprocessor-based system. Figure 10.10(b) shows the basic interfacing arrangement to the microprocessor system where the currents obtained from the secondary side of the current transformers are fed to analogue amplifiers for signal amplification. The currents are then sampled by sample-and-hold (SH) circuits and digitised by an analogue-to-digital convertor (ADC) before being presented to the microprocessor. The SH, multiplexer (MUX), A/D and D/A convertors are controlled by software. At the microprocessor, the data is processed using the

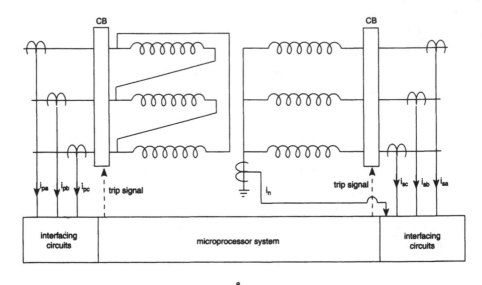

a

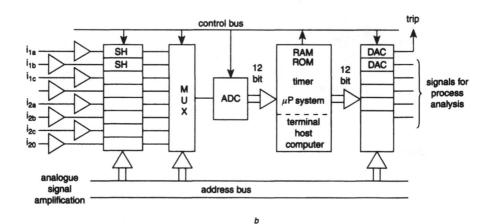

b

Figure 10.10   *Microprocessor based protective scheme for the protection of a three-phase, two-winding transformer*

(*a*) connection
(*b*) typical interfacing arrangement

CB = circuit breaker
SH = sample and hold
MUX = multiplexer
ADC = analogue to digital converter
DAC = digital to analogue converter

particular algorithm chosen. In the event of an internal fault a trip signal is sent through the D/A convertor to trip the circuit breakers on both sides of the transformer.

It will be noted that unlike conventional schemes, no interposing current transformers are required for phase rotation, as this is readily implemented by software.

10.8 References

1 KENNEDY, L.F., and HAYWORD, C.D.: 'Harmonic-current restrained relays for differential protection', AIEE 1938, **57**, pp. 262–266
2 HARDER, E.L., and MARTER, W.E.: 'Principles and practices of relaying in United States', AIEE, 1948, **67**, Pt II, pp. 1005–1022
3 MATTHEWS, C.A.: 'An improved transformer differential relay', *AIEE*, 1954, **73**, Pt III, pp. 645–650
4 ROCKEFELLER, C.D.: 'Fault protection with a digital computer', *IEEE Trans.* 1969, **PAS-88**, pp. 438–464
5 EINVALL, C.H., and LINDERS, J.R.: 'A three phase differential relay for transformer protection', *IEEE Trans.* 1975, **PAS-94**, pp. 1971–1978
6 HERMANTO, I., MURTY, Y.V.V., and RAHMAN, M.A.: 'A stand-alone digital protective relay for power transformers', *IEEE Trans.* 1991, **PWRD-6**, pp. 85–95
7 SCHWEITZER, E.O., LARSON, R.R., and FLECHSIG, A.J.: 'An efficient inrush current detection for digital computer relay protection of transformers', IEEE PES Winter Meeting, 1977, Paper A77510-1
8 LARSON, R.R., FLECHSIG, A.J., and SCHWEITZER, E.O.: 'The design and test of a digital relay for transformer protection', *IEEE Trans.*, 1979, **PAS-98**, pp. 795–804
9 DEGENS, A.J.: 'Algorithm for a digital transformer differential protection based on a least-squares curve fitting', *IEE Proc. C*, 1981, **128** (3), pp. 155–161
10 DEGENS, A.J.: 'Micro-processor-implemented digital filters for inrush current detection', *Int. J. Elec. Pow. & Energy Syst.*, 1982, **4**, pp. 196–205
11 RAHMAN, M.A., and JEYASURYA, B.: 'A state of the art review of a transformer protection algorithm'. *IEEE Trans.*, 1988, **PWRD-3**, pp. 534–544
12 RAHMAN, M.A., and DASH, P.K.: 'Fast algorithm for digital protection of power transformers', *IEE Proc. C*, 1982, **129** (2), pp. 79–85
13 RAHMAN, M.A., JEYASURYA, B., and GANGOPADHYAY, A.: 'Digital differential protection of power transformers based on Walsh functions', *Trans. CEA Engineering & Operating Div.* 1985, **24**, Paper 85-SP-149
14 JEYASURYA, B., and RAHMAN, M.A.: 'Application of Walsh functions for microprocessor based transformer protection', *IEEE Trans.*, 1985, **EMC-27**, pp. 221–225
15 THORP, J.S., and PHADKE, A.G.: 'A microprocessor based three phase transformer differential relay', *IEE Trans.* 1982, **PAS-94**, pp. 426–432
16 PHADKE, A.G., and THORP, J.S.: 'A new computer relay flux-restrained current differential relay for power transformer protection', *IEEE Trans.*, 1983, **PAS-102**, pp. 3624–3629
17 DEGENS, A.J., and LANGEDIJK, J.M.: 'Integral approach to the protection of power transformers by means of a microprocessor', *Elect. Pow. & Energy Syst.* 1985, **7**, pp. 37–47

Chapter 11

Digital line differential protection

11.1 Introduction

As power systems grow both in size and complexity, it becomes common to use long and heavily loaded two-terminal lines as well as multi-terminal and tapped lines [1–4]. This in turn has created difficult problems for their protection. Such problems include:

(i) Distance relay underreach:
     The infeed from another source to a fault point of a multi-terminal line causes distance relays to measure a higher impedance than the actual line impedance. This causes a delay in the tripping of the relay, which in turn may result in incorrect sequential tripping at other terminals.

(ii) Distance relay overreach:
     This problem arises when an external fault is fed from a number of terminals. This causes overreach, in which the distance relay measures a lower impedance than that of the fault loop. This can again cause false tripping and difficulty in determining appropriately the time selectivity between the protected and adjacent lines.

(iii) Effect of load current:
     Load current can cause an inaccurate impedance measurement by distance relays, which in turn can affect the integrity of the decision reached.

(iv) High resistance faults:
     A satisfactory operation of distance relays is not always achieved under highly resistive fault conditions, e.g. an earth fault through a tree. This type of fault is in many cases so similar to normal load conditions that the distance relay may fail to recognise it as a fault condition.

(v) Pilot-wire limitation:
     Pilot-wire relays can be applied successfully to overcome many of the problems described above. However, this is true only if the length of the line is not greater than typically 20 km because of the resistance and capacitance associated with the pilot wire. Various means of pilot compensation have been developed, but in general pilot-wire-based protection cannot be applied satisfactorily to many transmission circuits.

Recent developments in digital differential protection have been aimed at overcoming these limitations. These schemes can be broadly classified into current-based schemes and composite voltage- and current-based schemes.

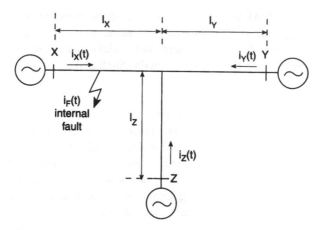

Figure 11.1 Basic three-terminal line configuration

11.2 Current-based differential schemes

11.2.1 Basic principles of line current differential protection [3–5]

The basic principle of line current differential protection used for two- and multi-terminal lines is essentially the same as the percentage-differential protection applied to transformers, which was discussed in Chapter 10.

Consider the three-terminal line shown in Figure 11.1. An instantaneous differential current signal $D(t)$ and a bias (or restrain) current signal $B(t)$ are typically formed using the instantaneous currents at the three ends such that

$$D(t) = i_x(t) + i_y(t) + i_z(t) \qquad (11.1)$$

$$B(t) = i_x(t) - i_y(t) - i_z(t) \qquad (11.2)$$

where $i_x(t)$, $i_y(t)$ and $i_z(t)$ are the instantaneous currents measured at ends X, Y and Z, respectively.

Under healthy conditions, ideally, the magnitude of the differential quantity $D(t)$ should be zero. In practice, it has a small value both under normal operating conditions and external faults, due to line charging current and other errors such as a mismatch of current transformers. However, once an internal fault occurs, the differential quantity $D(t)$ approximates to the fault current $i_F(t)$, i.e. the current flowing out of the fault point.

11.2.2 Frequency modulation current differential protective scheme [3, 4]

In frequency-modulation (FM) current-differential protective schemes, the instantaneous values of the currents at each end are frequency modulated and typically transferred to the other terminal(s) via a voice channel of a microwave communication network, as shown in Figure 11.2. Every terminal can be arranged to perform the functions of transmitting, receiving and initiating tripping signals when a fault is detected, and this is commonly called a master-

master-master (MMM) system (assuming a three-terminal line). The other way of arranging the system is to install a master in one terminal only, which receives current data from other terminals (slaves) and generates tripping signals which are sent to the slave terminals. Such a system is called a master-slave-slave (MSS) system.

The transmission of the instantaneous current values of each terminal is usually achieved by allocating a frequency band, from 300 to 3400 Hz for each current. The current at the secondary side of the CT at each terminal is converted into a voltage, which in turn is converted into a frequency signal in the appropriate voice band before it is transmitted to remote terminals. Either a microwave communication link or optical fibres can be used to transmit the necessary signals. At the receiving end, the current signal is recovered by conversion of the frequency-modulated signal to a voltage signal.

FM relay characteristics
Although the FM differential relay relies fundamentally on the percentage differential principle, its percentage restraining characteristics are often slightly different from the conventional type applied to transformer protection. Figure 11.3 shows the basic characteristics of a typical percentage FM differential relay. It will be seen that the characteristics consist of two sections, AB and BC. The slope of section AB, the small current region, is chosen to permit the detection of small internal fault currents under the existence of heavy load conditions. It is also effective for detecting small internal fault currents under high arc resistance conditions. On the other hand, the slope of section BC is such that it can deal with heavy internal fault conditions. This section is also effective in preventing false tripping during external fault conditions, which cause CT saturation and thus generate significant differential-current signal components.

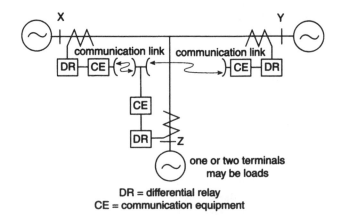

Figure 11.2   Basic construction of FM relay system

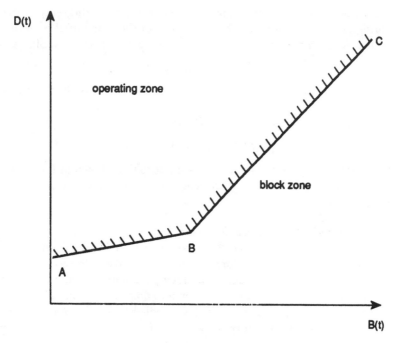

Figure 11.3   Percentage restraining characteristic of an FM current-differential relay

It is important to note that differential and bias signals are commonly derived separately for each phase of the line. In this way, many of the problems formerly associated with the summation of phase quantities are overcome in modern digital line differential protection arrangements.

11.2.3   Modal current based protection scheme [5]

Operating principles and tripping criteria
The operating principles of this relay also depend on the percentage-differential current concept. However, the differential quantity $D(t)$ and the bias quantity $B(t)$ are derived using the instantaneous values of modal currents at the line terminals. Let us first consider these principles using total time variations for a three-terminal line. With reference to Figure 11.1, the two quantities $D(t)$ and $B(t)$ at the master end are given by eqns. 11.1 and 11.2, respectively. It will be noted that, under temporary two-terminal operation involving transmission between ends Y and Z, or conditions where there is a complete loss of infeed at end X, the differential signal $D(t)$ and the bias signal $B(t)$ will be equal in magnitude. This is of no consequence for internal faults, but with external faults causing an abnormally high degree of CT saturation at ends Y and Z, it could cause sufficient unbiased differential current to be processed and thereby lead to relay instability. This problem can be overcome by adding a logic check within the tripping decision process, as will be explained later.

In its simplest form, the tripping criterion adopted by the relay is that the tripping signal is initiated when the magnitude of the differential quantity $D(t)$ exceeds that of the bias quantity $B(t)$ by a certain predefined threshold value K_S:

$$|D(t)| - K_B|B(t)| \geqslant K_S \qquad (11.3)$$

However, to avoid the above-mentioned problem, a modified process described by eqn. 11.4 is used. In this case the differential current component $D(t)$ is checked against a time-variant threshold signal $S(t)$:

$$|D(t)| \geqslant S(t) = K_S + K_B|B(t)| \qquad (11.4)$$

The basic arrangement typically used is similar to that shown in Figure 11.2 for an MMM scheme, although MSS schemes are common when using this approach.

Modal components of differential and bias quantities

As previously discussed in Chapter 8, assuming an ideally transposed line, phase quantities can be transformed into what are referred to as modal quantities. This has the advantage of producing relay measurements that do not require complex transformations, and therefore do not delay signal components in their derivation. The first modal component of differential and bias quantities $D_1(t)$ and $B_1(t)$ is formed as the difference of the 'a' and 'c' phase currents, such that

$$D_1(t) = [i_{ax}(t) - i_{cx}(t)] + [i_{ay}(t) - i_{cy}(t)] + [i_{az}(t) - i_{cz}(t)]$$
$$B_1(t) = [i_{ax}(t) - i_{cx}(t)] - [i_{ay}(t) - i_{cy}(t)] - [i_{az}(t) - i_{cz}t)] \qquad (11.5)$$

The second modal component based signals $D_2(t)$ and $B_2(t)$ are commonly found from the difference between 'a' and 'b' phase currents:

$$D_2(t) = [i_{ax}(t) - i_{bx}(t)] + [i_{ay}(t) - i_{by}(t)] + [i_{az}(t) - i_{bz}(t)]$$
$$B_2(t) = [i_{ax}(t) - i_{bx}(t)] - [i_{ay}(t) - i_{by}(t)] - [i_{az}(t) - i_{bz}(t)] \qquad (11.6)$$

where i_{ax}, i_{bx}, i_{cx} are the 'a', 'b', 'c' phase currents at end X, i_{ay}, i_{by}, i_{cy} are the phase currents at end Y, and i_{az}, i_{bz}, i_{cz} are the corresponding currents at end Z.

Since in practice there are no fault conditions that can simultaneously give rise to zero-valued modal current signals, tripping is initiated for internal faults by at least one modal channel. This form of signal compression reduces the communication channel requirement from three (for separate phase-by-phase comparison) to two.

Trip-decision logic process

A typical trip-decision logic process can best be explained by reference to Figure 11.4, where (*a*) shows a flow diagram of the decision process algorithm. Figure 11.4(*b*) shows typical differential-current waveforms for external and internal fault conditions. The former is small and comprises mainly spill current, which is predominantly high frequency, whereas the latter is large and predominantly power frequency. To simplify the explanation, the pick-up level is taken as $\pm K_S$

instead of the time variant signal $\pm S(t)$ (see eqn. 11.4), i.e. K_B is assumed to be zero.

Let us first consider the external fault waveform. It can be seen that the magnitudes of the four samples 2–5 are above the pick-up levels $\pm K_S$. Thus,

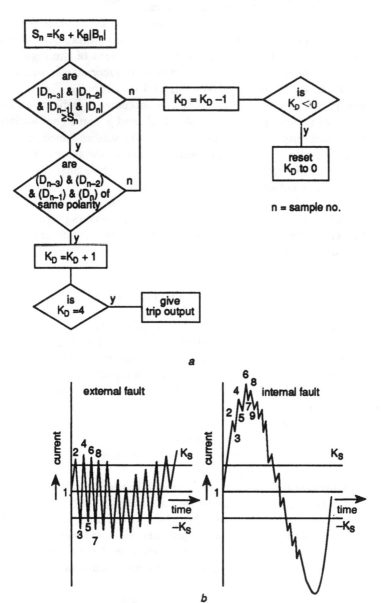

a

b

Figure 11.4   Typical extended digital decision logic process [5]

(*a*) decision logic algorithm
(*b*) typical waveforms

after the completion of the first operation of the logic, the decision counter K_D is set for an up count. However, since the polarities of these samples alternate between positive and negative, as shown on the waveform, the second operation of the logic results in a down count. Consequently K_D, for the waveform under consideration, stays close to zero at all times.

Consider now the internal fault waveform. It can be seen that once the differential current has exceeded the pick-up level, it stays above that level for an appreciable time. This means the two operations of the logic related to comparisons of magnitude and polarities of four successive samples at a time will allow K_D to attain the required value of 4 very rapidly. For this particular waveform, it can be seen that samples 2–8 successively indicate an up count. The criteria of adopting a four sample check and a trip signal initiation at a decision counter output of 4 is based on extensive studies aimed at maximising relay stability under external fault conditions and at the same time maximising the sensitivity to internal faults. The adoption of the bias quantity $K_B|B(t)|$ (see eqn. 11.4) further enhances stability under healthy conditions without significantly affecting performance for internal faults. In essence, the trip decision logic described makes use of further features of the relaying signals to enhance the integrity of the decision asserted under onerous conditions, which may otherwise cause relay malfunction.

11.3 Composite voltage- and current-based scheme [6, 7]

Although current differential schemes are relatively fast in terms of fault-clearance times, they have the common problem of becoming insensitive to low levels of fault current because of their total dependence on current. This arises from the standard practice of providing a bias signal proportional to through current to ensure security under conditions where a significant differential signal is generated under healthy conditions, e.g. due to shunt capacitance charging currents or current transformer errors. A second problem is that certain system operating conditions may result in the difference between the differential and restraint signals being small for internal faults, which means a reduction in dependability of the protection.

To overcome these problems, an approach has recently been adopted which involves deriving differential signals that are functions of both the voltages and currents measured at each end of either plain or teed lines. Consequently, a bias signal to compensate for capacitance spill currents is no longer required. This in turn increases the relay sensitivity to a much lower level of fault currents than provided by schemes that are totally dependent on current measurement.

11.3.1 Basic operating principles

The basic operating principle relies on deriving signals proportional to the instantaneous values of modal voltages and currents at the three ends of the line. Ideally the sum of these signals should be zero under healthy conditions

Figure 11.5   Basic two-terminal line configuration giving current-reference directions

but significantly finite under internal fault conditions. Thus, for a three-terminal line such as that shown in Figure 11.1 we have

$$S_X^{(k)} + S_Y^{(k)} + S_Z^{(k)} \begin{cases} = 0 \text{ healthy conditions} \\ \neq 0 \text{ internal fault conditions} \end{cases} \quad k = 0, 1, 2 \quad (11.7)$$

For a two-terminal line the signal S_Z is equal to zero and the above equation reduces to

$$S_X^{(k)} + S_Y^{(k)} \begin{cases} = 0 \text{ healthy conditions} \\ \neq 0 \text{ internal fault conditions} \end{cases} \quad k = 0, 1, 2 \quad (11.8)$$

where S_X, S_Y and S_Z are the terminal signals at ends X, Y and Z, and k is the mode number.

11.3.2 Formation of terminal signals

11.3.2.1 Two-terminal lines [6]
Assuming a distributed-parameter, transposed three-phase transmission line, the modal voltage and current relationships in the frequency domain can be described by eqn. 8.30:

$$\frac{d^2 V^{(k)}(\omega)}{dx^2} = (\gamma^{(k)}(\omega))^2 V^{(k)}(\omega), \frac{d^2 I^{(k)}(\omega)}{dx^2} = (\gamma^{(k)}(\omega))^2 I^{(k)}, k = 0, 1, 2 \quad (11.9)$$

As previously explained in Chapter 8, the above equation shows that wave propagation in a three-phase line can thus be considered in terms of three independent components, each possessing its own modal propagation constant $\gamma^{(k)}(\omega)$ and associated modal surge impedance $Z_0^{(k)}(\omega)$, where $k = 0$, 1 and 2.

The solutions of the voltage and current differential equations shown in eqn. 11.9 take the form

$$V^{(k)}(\omega) = K_1 \, e^{-\gamma^{(k)}x} + K_2 \, e^{\gamma^{(k)}x} \quad (11.10)$$

and

$$I^{(k)}(\omega) = [K_1 \, e^{-\gamma^{(k)}x} - K_2 e^{\gamma^{(k)}x}]/Z_0^{(k)}$$

where K_1 and K_2 are arbitrary constants and x is an arbitrary length of line.

In the above equation, the frequency variable ω has been dropped from the propagation constant $\gamma^{(k)}$ in order to simplify the notation.

The arbitrary constants K_1 and K_2 can be determined from a knowledge of the boundary conditions for a particular system. Consider the two terminal line shown in Figure 11.5, which has the following boundary conditions:

$$x = \begin{cases} 0 & \text{at end X} \\ l & \text{at end Y} \end{cases} \quad (11.11)$$

where l is the length of the line.

By substituting eqn. 11.11 into eqn. 11.10, we obtain the voltage and current at end X:

$$V_X^{(k)}(\omega) = (K_1 + K_2)$$
$$I_X^{(k)}(\omega) = (K_1 - K_2)/Z_0^{(k)}(\omega) \qquad k = 0, 1, 2 \qquad (11.12)$$

and those relating to end Y will be

$$V_Y^{(k)}(\omega) = K_1\, e^{-\gamma^{(k)}l} + K_2\, e^{\gamma^{(k)}l}$$
$$I_Y^{(k)}(\omega) = (K_1\, e^{-\gamma^{(k)}l} - K_2\, e^{\gamma^{(k)}l})/Z_0^{(k)} \qquad k = 0, 1, 2 \qquad (11.13)$$

By substituting the values of K_1 and K_2 obtained from eqns. 11.12 into eqns. 11.13, we finally obtain

$$V_Y^{(k)}(\omega) + Z_0^{(k)}(\omega)I_Y^{(k)}(\omega) = e^{-\gamma^{(k)}l}[V_X^{(k)}(\omega) + Z_0^{(k)}(\omega)I_X^{(k)}(\omega)] \; k = 0, 1, 2 \qquad (11.14)$$

By using the current and voltage reference directions shown in Figure 11.5, i.e. by reversing the direction of the current I_Y at end Y, eqn. 11.14 becomes

$$V_Y^{(k)}(\omega) - Z_0(\omega)I_Y^{(k)}(\omega) = e^{-\gamma l}\,[V_X^{(k)}(\omega) + Z_0^{(k)}(\omega)I_Y^{(k)}(\omega)] \quad k = 0, 1, 2 \qquad (11.15)$$

or $S_X^{(k)} + S_Y^{(k)} = 0$ $\qquad\qquad\qquad\qquad\qquad k = 0, 1, 2 \qquad (11.16)$

where $S_X(k)$ is equal to the right-hand side of eqn. 11.15 and $S_Y^{(k)}$ is equal to the left-hand side negated.

11.3.2.2 Three-terminal lines

The basic theory laid down above for the two-terminal lines can be easily extended to include teed (or three-terminal) feeders. Consider again Figure 11.1, which shows that a three-terminal line basically consists of three separate line sections. These are sections XT, YT and ZT. For each of these sections an equation similar to eqn. 11.14 can be derived in terms of the current and voltage at the line ends. When the resulting equations are combined together (and applying the fact that the sum of currents at the tee point T is zero), we obtain

$$S_X^{(k)} + S_Y^{(k)} + S_Z^{(k)} = 0 \qquad k = 0, 1, 2 \qquad (11.17)$$

where

$$S_X^{(k)} = A_X V_X^{(k)}(\omega) + B_X Z_0^{(k)}(\omega)I_X^{(k)}(\omega)$$
$$S_Y^{(k)} = (A_Y - B_Y)V_Y^{(k)}(\omega) + (A_Y + B_Y)Z_0^{(k)}(\omega)I_Y^{(k)}(\omega)$$
$$S_Z^{(k)} = (A_Z - B_Z)V_Z^{(k)}(\omega) + (A_Z + B_Z)Z_0^{(k)}(\omega)I_Z^{(k)}(\omega)$$
$$A_X = e^{-\gamma^{(k)}l_X} - 1, \; B_X = e^{-2\gamma^{(k)}l_X} + 1$$
$$A_Y = e^{-[\gamma^{(k)}(l_X + l_Y)]}, \; B_Y = e^{-[\gamma^{(k)}(l_X - l_Y)]}$$
$$A_Z = e^{-[\gamma^{(k)}(l_X + l_Z)]}, \; B_Z = e^{-[\gamma^{(k)}l_X - l_Z)]}$$

l_X, l_Y, l_Z are the lengths of sections XT, YT and ZT respectively.

11.3.2.3 Trip-decision logic

It is seen from the foregoing that, in theory, the signals at the ends of a line (whether of the two- or three-terminals type) should sum to zero under all healthy conditions. However, because of quantisation, transducer errors etc, it becomes necessary to apply a small threshold T_h to the sum signal of eqn 11.17. Thus in practice the simplest relay tripping decision would be

$$|S_X^{(k)} + S_Y^{(k)} + S_Z^{(k)}| > T_h \qquad k = 0, 1, 2 \qquad (11.18)$$

The above equation applies to a three-terminal application. In the case of a two-terminal line $S_Z^{(k)}$ becomes zero.

It will be noted that the tripping decision described by eqn. 11.18 is fairly simple compared with that used in the previously described current differential schemes. In consequence, a relatively simple trip-decision logic process can be used to provide the necessary degree of dependability and sensitivity. The simplest criterion that provides an adequate performance for most protective applications involves the initiation of a trip signal if two consecutive samples of the signal summation of eqn. 11.18 exceed the fixed threshold setting T_h.

11.4 References

1 AIEE Working Group of the Line Relay Protection Sub-Committee: 'Protection of multiterminal and tapped lines', AIEE Fall Meeting, October 1960, Paper CP60-1274
2 'IEEE Study Committee report on protection aspects of multi-terminal lines', IEEE Report 79, THOO56-2-PWR, 1979
3 KITAGAWA, M. *et al.*: 'Newly developed FM current-differential carrier relaying system and its field experience', IEEE PES Winter Meeting, 1978, Paper F78, 291–297
4 YAMAURA, M., MASUI, M., and OKITA, Y.: 'FM current differential carrier relaying', Developments in power system protection, (IEE Conf. Publ. 185, 1980, pp. 156–160)
5 AGGARWAL, R.K., and JOHNS, A.T.: 'The development of a new high speed three-terminal line protection scheme', IEEE/PES Summer Meeting, 1985, Paper 85 SM 3200-3207
6 AGGARWAL, R.K., and JOHNS, A.T.: 'A differential line protection scheme for power systems based on composite voltage and current measurements', IEEE/PES Winter Meeting, 1989, Paper 89WM 053-0 PWRD
7 AGGARWAL, R.K., and JOHNS, A.T.: 'New approach to teed feeder protection using composite current and voltage signal comparison', Developments in power system protection (IEE Conf. Publ. 302, 1989, pp. 125–129)

Index

CPSIA information can be obtained
at www.ICGtesting.com
Printed in the USA
JSHW051427050721
16597JS00001B/61